Reactivity and Structure
Concepts in Organic Chemistry

Volume 30

Editors:

Klaus Hafner
Charles W. Rees
Barry M. Trost

Jean-Marie Lehn
P. von Ragué Schleyer
Rudolf Zahradník

Michinori Ōki

The Chemistry of Rotational Isomers

With 3 Figures and 42 Tables

Springer-Verlag
Berlin Heidelberg NewYork
London Paris Tokyo
Hong Kong Barcelona Budapest

Professor Michinori Ōki
Department of Chemistry
Okayama University of Science
Okayama 700, Japan

Library of congress Cataloging in Publication Data
Ōki, Michinori 1928.
The chemistry of rotational isomers / Michinori Ōki, p. cm. (Reactivity and structure ; v. 30),
Includes bibliographical references and index.

1. Isomerism. II. Title. III. Series.
QD471.038   1993   541.2'252--dc20   92-37400   CIP

ISBN 978-3-642-51026-7   ISBN 978-3-642-51024-3 (eBook)
DOI 10.1007/978-3-642-51024-3

© Springer-Verlag Berlin Heidelberg 1993
Softcover reprint of the hardcover 1st edition 1993

Typesetting: Macmillan India Ltd., Bangalore, India
2152/3020-5 4 3 2 1 0

List of Editors

Preface

The concept of internal rotation was developed as early as 1875 but was never thought meaningful in organic chemistry until recent times. Indeed, rotation about a single bond in organic compounds has been assumed for a long time to be free and it was only about 50 years ago when the concept of restricted rotation was established from physical measurements. Even after learning of the presence of rotational isomers by spectroscopic measurements, organic chemists tended to neglect the presence of such isomers. The presence of atropisomers in biphenyls has been a very special case and they give differences only under chiral conditions: usually we distinguish diastereomers but not enantiomers.

However, knowledge of rotational isomers has been accumulating rapidly in recent years. Even organic chemists cannot ignore the difference in rotational isomers today. This is, on one hand, caused by advances in spectroscopy and, on the other, by advances in asymmetric syntheses.

We were fortunate to be able to work on two series of compounds that give stable rotational isomers at room temperature and to be able to examine the differences in their reactivities. This book is the compilation of our recent work in this field together with some historic background and related areas.

Chapter 1 describes the historical concept of "free rotation" about a single bond, development of the concept of restricted rotation and rotational isomers, and isolation of stable rotational isomers at room temperature. This chapter also deals with the concepts which still prevail among some chemists today but should be changed: internal compensation and chemical properties of diastereomeric rotational isomers.

Chapter 2 describes factors that affect populations of rotational isomers. Since the two series that give stable rotational isomers at room temperature feature the close proximity of two groups in a molecule in the ground state, it is possible to observe very weak molecular interactions, that cannot be otherwise detected, as unusual population ratios of rotational isomers. These series of compounds serve as excellent probes for detecting weak molecular interactions which are theoretically possible but are difficult to observe.

Chapter 3 describes factors that affect the height of the barrier to rotation, after briefly mentioning about the method for determining the barrier height. An interesting point here is that a bulky substituent does not necessarily increase the height of the barrier. Emphasis is put on the fact that the barrier height is the difference in energies between the ground state and the transition state for rotation.

Chapter 4 describes the differences in the reactivities between rotational isomers. It is also pointed out that the diastereotopic methyls such as those in a *tert*-butyl group can show different reactivities. The origin of the difference in reactivities is, of course, intra(and/or inter) molecular interactions which accelerate or retard formation of reaction intermediates or transition states. Readers will notice that there is even a case in which rotational isomers react via different mechanisms from each other under the same conditions.

It will be a great pleasure for the author if this book helps organic chemists realize that a mixture of diastereomeric rotational isomers should be considered a mixture of differently reacting isomers in the laboratory and, if they are considering reactions under chiral conditions, any of the rotational isomers can behave differently. It is also hoped that this book causes further development of chemistry in the field of rotational isomers by motivating young chemists.

There is one thing which should be mentioned here concerning the presentation of this book. Showing stereochemical structures is very space-demanding. Because we are dealing with rotational isomers in this book, it is normally necessary to show three or more stereochemical drawings for one compound which can sometimes fill a page of printing. To avoid this situation, after a time, only one isomer of the two possible enantiomers is shown. Therefore, the readers are advised not to take these structures as if they are separated into optical isomers but take them as showing the possible pair of enantiomers.

The author owes a great deal to former and present associates in The University of Tokyo and Okayama University of Science, without whose enthusiasm this kind of chemistry could not have been explored. The author also wishes to thank Professor Michael Hanack at Tübingen, Professor Horst Kessler in München, and Professor Jan Sandström in Lund for their reading of the manuscript.

Okayama, Japan, May, 1992   Michinori Ōki

Table of Contents

1 Introduction

1.1 The "Free Rotation" Concept

It was a historical breakthrough when van't Hoff [1] and LeBel [2] were able to explain the existence of enantiomers in carbon compounds by assuming the tetrahedral arrangement of four ligands about the carbon atom. Although it took sometime for the theory to be accepted by the chemical world, it became the general concept that the carbon atoms take on a tetrahedral spatial arrangement and the organic molecules should be treated as three dimensional objects. The reason for this is the easy understanding of the optical activity of isomeric molecules which differ only in the direction of optical rotation but possess identical properties in other chemical and physical aspects.

The feature of the optically active compounds was existence of at least one carbon atom that carry four substituents, any two of which were not identical. This carbon atom was called an asymmetric carbon atom and today it is better known as a chiral center. When one writes the tetrahedral arrangement of the four substituents around a carbon atom, it becomes clear that its mirror image is not superposable with it, as can be understood by the following scheme.

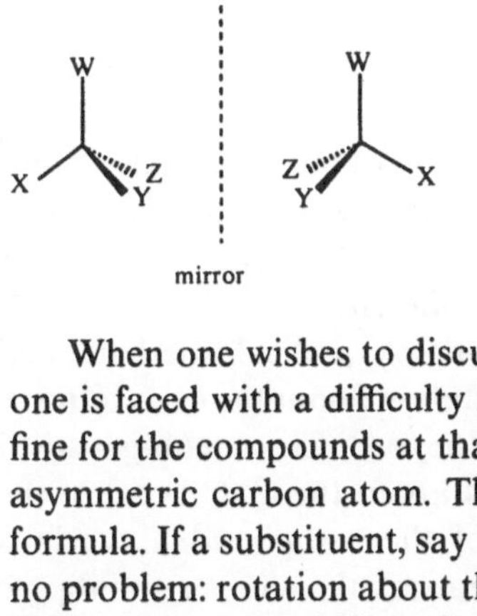

mirror

When one wishes to discuss the stereochemistry of two carbon compounds, one is faced with a difficulty immediately, although the tetrahedron theory was fine for the compounds at that time, when one considers stereochemistry of one asymmetric carbon atom. This is best represented by the following structural formula. If a substituent, say W of the foregoing compound, is spherical, there is no problem: rotation about the C–W bond causes no isomer. It is also true that, if the symmetry of the substituent is C_{3n} and if one accepts that the energy minimum is the place where A and B reside at the middle point of the substituents on the other carbon atom, as is accepted today, then rotation about the C–C bond causes no isomer. But as soon as the symmetry of W is lower than that, one has to worry about the presence of other kinds of isomers which was not recognized in the time of van't Hoff and LeBel: namely the compounds which are shown by the above scheme are different and are formed by mere rotation about the C–W bond.

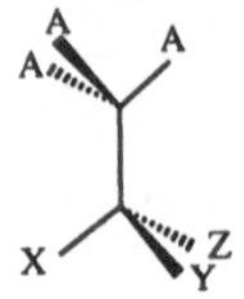

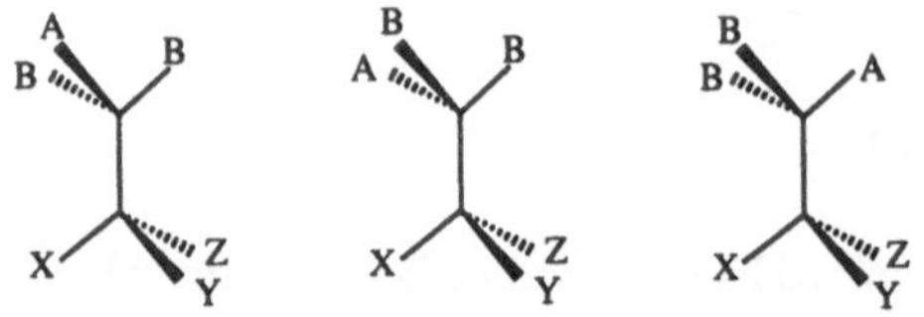

Since the presence of such isomers was not known at the time of the tetrahedron theory, van't Hoff was forced to assume easy rotation about a carbon-to-carbon single bond to accommodate the fact that there could be no such isomers: if the parts of the molecule were freely rotating about the C–W single bond, it would not be possible to recognize any isomer due to such rotation [3]. Thus the concept of free rotation was implied by van't Hoff and became common among organic chemists at least for some time. Free rotation is used today in a different context. Therefore, if one wishes to use free rotation in the classical sense, it is advisable to mention that point, namely one only refers to the fact that rotational isomers are not isolated by the classical means. This point is dealt with further in Sect. 1.2.

1.2 Recognition of Rotational Isomers

There was no problem when organic chemists were working at the laboratory bench, where the time scale is to the order of hours because the rotation about a single bond is much faster than the time scale. However, when spectroscopy was introduced into chemistry, things became a little different.

Mizushima et al. measured the Raman scattering of 1,2-dichloroethane and found that the results were not explained by free rotation about the C–C bond of the compound [4]. The spectra showed four lines due to C–Cl stretching instead of two, which was expected from the free rotation about the C–C bond of the compound. The dipole moment measurement of the compound at various temperatures indicated that the lower the temperature, the smaller the dipole moment. This was also not in conformity with the idea of free rotation about the single bond in the compound. The results were interpreted in the following way.

If we assume there are three rotational isomers as shown on the next page, there will be four modes of C–Cl stretching, two symmetric and two antisymmetric. Furthermore, the isomer which possesses the two C–Cl bonds at a 180° arrangement will be nonpolar with zero dipole moment. If therefore, the isomer is more stable, it is natural to expect that the isomer will increase in its population when temperature is lowered, thus making the dipole moment of the whole smaller. These rotational isomers must be rapidly equilibrating because they were not isolated at room temperature.

Pitzer and his coworkers worked on the entropy of gases. They found that the entropy of ethane was not reproduced by statistical mechanics if free rotation about the C–C bond was assumed, nor was it possible to do so for the molecules that possess rigid structure. However, when it was assumed that there were stable and unstable points during the rotation about the bond in question, of which potential energy could be expressed by sinusoidal curves, then it was possible to reproduce the results by statistical mechanics [5].

Thus the interpretation of the experimental results is as follows. During the course of rotation about the C–C bond in ethane, two hydrogen atoms which are connected to different carbon atoms will take an arrangement which is depicted by the Newman projection as eclipsed and in the next instance they will be farthest apart, by being in an arrangement that is depicted by the staggered form in the same projection. If two hydrogens are close to each other, the potential energy will be heightened because of various repulsive forces, whereas the energy will be minimum when the two hydrogens are apart as far as possible in the molecule. Then there will be three maxima and three minima of potential energy during the course of rotation of 360° in ethane. Assuming these potentials, the barrier to rotation about the C–C bond in ethane was obtained as ca. 3 kcal/mol which gave the best fit of the calculated values with the observed.

eclipsed   staggered

This barrier to rotation has an important meaning. At room temperature, the molecules in the gas phase possess ca. 0.6 kcal/mol kinetic energy. The barrier to rotation in ethane surpasses this value. That is, the internal rotation in ethane takes place slowly at room temperature, it remains as staggered forms longer than as eclipsed forms. This is by no means free rotation. Therefore this type of rotation is called restricted rotation [6]. Then strictly speaking, free rotation is the internal rotation where no activation energy is required. There are several examples known as free rotation today such as the rotation of a methyl group in acetylene derivatives in the gas phase [7]. One has to recognize that the "free rotation" in the classical sense is now very often restricted rotation.

It is easily recognized that butane can possess three stable staggered forms, as 1,2-dichloroethane does. They can be depicted by the Newman projection as in the following scheme which also shows the equilibration among these isomers. The two forms of the three are enantiomers, because their mirror images are not superposable with the original bodies. They are shown in the scheme as + sc and − sc. The remaining one is a *meso* form because it possesses a symmetry plane in the molecule and its mirror image is superposable to the original body according to the classical theory. According to the repulsion theory, the *meso* form which is now called the *ap* form must be more stable than the enantiomeric *sc* forms, because in the latter two bulky methyl groups are located closely. Since the *ap* and the *sc* forms are isomers, they are called rotational isomers or conformational isomers with each other. According to the current terminology, the *ap* and *sc* isomers are diastereomeric with each other. Their physical properties as well as chemical properties must be different even under achiral conditions.

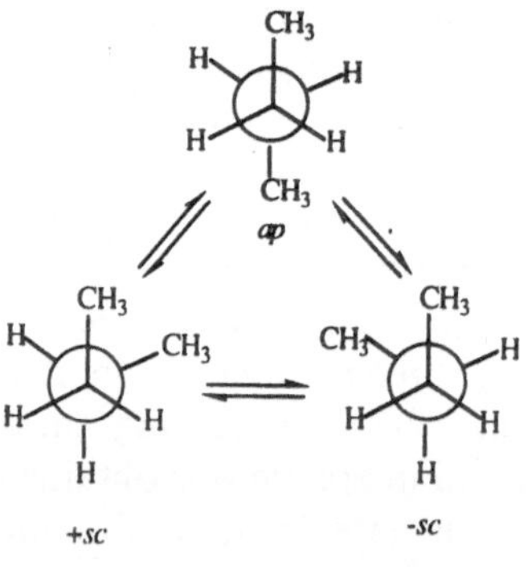

These rotational isomers are usually not recognized as such when one discusses properties of the molecules on a slow time scale. For example, NMR spectroscopy cannot recognize many rotational isomers that are recognized by infrared or Raman spectroscopy, because the time scale of the former is much slower than the latter two. In order to describe these cases, the term "residual isomerism" was coined [8].

Conformations of olefins about the C–C bond which connects the double bond to an sp^3-hybridized carbon atom became interested much later than those of ethane derivatives. Propene is known to be stable in a form (*sp*-**1-1**) in which the double bond eclipses one of the C–H bonds rather than in a form ($\pm$ *sc*-**1-1**) in which one of the C–H bonds of the methyl group eclipses the C–H single bond of the olefinic part by microwave spectroscopy [9]. This is now confirmed by a sophisticated calculation and the cause of the stability is said to be orbital interactions, stronger attraction due to $\pi_{Me}-\pi^*_{CC}$ and $\pi_{CC}-\pi^*_{Me}$ and weaker repulsion due to $\pi_{Me}-\pi_{CC}$ in the eclipsed form than in other conformations [10].

By taking such conformations, 1-butene can exist in three rotationally isomeric forms (*sp*-**1-2**, $+$ *ac*-**1-2** and $-$ *ac*-**1-2**) as shown below. A microwave spectrum indicates that the methyl-eclipsing form (*sp*-**1-2**) is as stable as the others ($+$ *ac*-**1-2** and $-$ *ac*-**1-2**) [11].

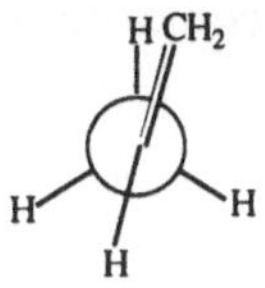

sp-**1-1** 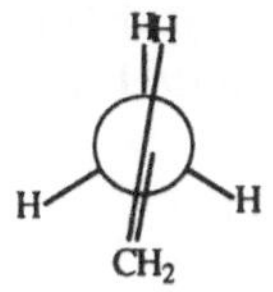   ±sc-**1-1**

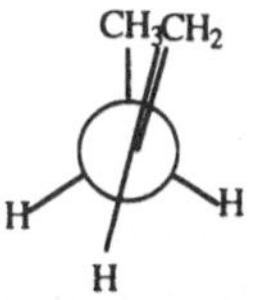

sp-**1-2** 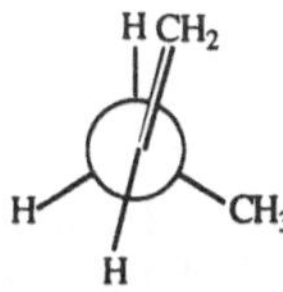   +ac-**1-2** 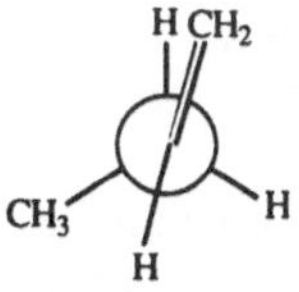   -ac-**1-2**

1.3 Atropisomers

In 1922, optical isomers of a biphenyl derivative (**1-3**) were isolated [12], the isolation being possible due to the fact that the internal rotation about the pivot bond (6–6′ bond) is frozen. For this phenomenon the term "atropisomerism" was coined [13].

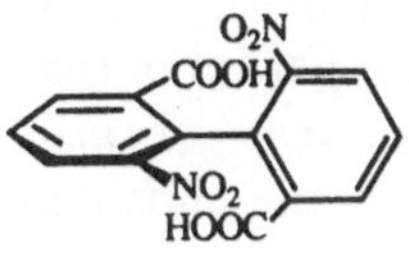

(R)-**1-3** 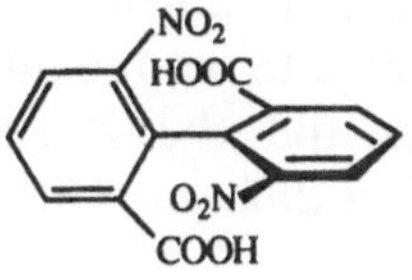   (S)-**1-3**

mirror

In earlier days, atropisomerism was used by organic chemists and rotational isomerism by physical chemists [14]. However, since "atrop" means "not rotating", atropisomerism is now used in a much broader sense than originally meant. It can of course be used in the biphenyl series but has been extended in usage so as to mean that the rotational isomers are detected as a result of NMR spectroscopy at low temperatures, though those isomers are not isolated at room temperature. Similarly, some authors use this term to mean that rotational isomers were detected by infrared spectra. In this sense, the rotational isomerism and the atropisomerism are the same. One has to be cautious and define the methodology with which one is dealing and to mention the temperature of which the phenomenon is observed when the terms rotational isomerism or atropisomerism are used.

A good example of this class is the isolation of chlorocyclohexane (**1-4**) in its equatorial form [15]. The barrier to inversion of the cyclohexane ring is known to be ca. 10 kcal/mol [16]; that is of course too low to permit isolation of the isomers at room temperature. At $-150\,°C$, the ring inversion is very slow and the equatorial form, which is more stable and thus richer in conformational population than the axial form, can be isolated in pure form.

equatorial   axial

1-4

Though the term atropisomerism can be used for both biphenyl (or 1,3-butadiene type) and ethane or propene derivatives, there is one point which should not be overlooked. That is, it is often cited that the restricted rotation about the sp^2-sp^2 bond is a phenomenon that affords only enantiomers, whereas that about the sp^2-sp^3 or sp^3-sp^3 bond gives both enantiomers and diastereomers. However, one can consider the rotational isomers of a biphenyl derivative which has the R (or S) stereochemistry. They are diastereomers as are shown by the next models (**1-5**). Since they are not usually separable by the classical means, due to the low barriers separating them, the chemistry of these isomers are seldom explored. However, it is only the matter of difficulty in recognizing these isomers in the case of biphenyl and 1,3-dienes.

If the time scale of the experiment is different from that of the normal laboratory bench, it is of course possible to see the different properties of these diastereomers. For example, the E and the Z isomers of o,o'- or m,m'-difluorobiphenyl were shown to exhibit different ESR spectra of the phosphorescent triplet states at 77 K [17].

Z-isomer   E-isomer

1-5

1.4 Isolation of Rotational Isomers

According to the present day explanation, the Raman spectra (or infrared absorption spectra) can distinguish two isomers which are separated by a barrier of ca. 3 kcal/mol or less due to the frequency difference of the two signals concerned. Therefore, the time scale of the vibration spectra are of the order of $10^{11} s^{-1}$. This time scale is fine enough to detect practically all the rotational isomers as if frozen. This means however that it is not possible to estimate the height of the barrier which separates two rotational isomers by vibration spectroscopy.

The barrier to rotation can be determined by far-infrared spectroscopy or microwave spectroscopy [18], if a molecule in question is rather simple and the barrier height is low. Therefore, the barriers to rotation in simple molecules have been obtained by these techniques. By contrast, NMR spectroscopy can cover the barrier to rotation ranging from 5 kcal/mol to 25 kcal/mol due to its time scale (life time $10^0 - 10^{-3} s^1$) [19]. Futhermore, it is possible to apply NMR spectroscopy to fairly complicated organic molecules by choosing appropriate probes which are borne by various substituents for the determination of barriers to rotation. Thus this methodology has become a choice of many organic chemists in studying and determining the barrier to large amplitude motions within a molecule. Indeed, the barrier of 25 kcal/mol is high enough to permit one to isolate rotational isomers at room temperature, thus giving chemists directions in selecting organic compounds for isolation of rotational isomers.

1.4.1 Factors Affecting Barrier Heights

The simple treatment of the structure of the transition state for rotation in ethane takes into consideration the repulsive forces between the electrons that belong to nonbonded atoms. According to this concept, the barrier to rotation will be enhanced if one introduces a substituent in the ethane molecule.

As expected, butane possesses a higher barrier to rotation about the C_2-C_3 bond than ethane, the barrier being estimated to be ca. 4.5 kcal/mol [20]. The natural consequence is to suppose that the bulkier the substituent the higher the barrier.

This is partly materialized because 2,3,3-trimethyl-2-butyl derivatives (1-6) show a much higher barrier to rotation of about 10 kcal/mol than butane [21]. Even enantiomers of 1,1,2,2-tetra-*tert*-butylethane (1-7) are reported to be separable, though attempted estimation of the barrier to rotation failed because of decomposition [22]. A similar compound which carries adamantyl groups instead of the *tert*-butyls in compound **1-7** can also be separated into rotational isomers [23].

This consideration is in accord with the fact that, in the well-known examples of biphenyl derivatives, the barrier to rotation is higher as the bulkiness of the 2,6-substituents increases [24]. However, things are not so simple. The barriers to rotation about the C–C bond in 2,2,3,3-tetrahalobutanes (**1-8**) were found to be as high as 16 kcal/mol [25], an increase by ca. 6 kcal/mol from the 2,3,3-trimethyl-2-butyl derivatives, irrespective of the fact that a methyl group is larger than a chloro and is of about the same size with a bromo substituent. The cause for this increase is attributed to the dipolar interactions at the transition state for rotation, because polar carbon-to-halogen bonds are in a parallel arrangement thereby increasing the barrier height.

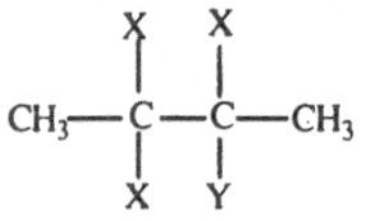

$$CH_3 \overset{\overset{X}{|}}{\underset{\underset{X}{|}}{C}} \overset{\overset{X}{|}}{\underset{\underset{Y}{|}}{C}} CH_3$$

1-8

However, there is another point about the energy of the ground state that deserves mention. In the biphenyl system, the introduction of a bulky substituent into the 2- or 6-position does not influence the ground state energy, because the phenyl group is planar, but does influence the transition state energy for rotation, because in the transition state those substituents interact strongly with the biphenyl skeleton. This can be clearly understood from the models (**1-9**) shown below and is characteristic of the cases of restricted rotation about an sp^2–sp^2 bond. However, in the case of 2,3,3-trimethyl-2-butyl derivatives, the introduction of the substituent influences not only the transition state for rotation, but also the ground state, as can be seen from the following structure (**1-10**): by congesting the molecule, the ground state energy is raised. The difference between the energies of the ground state and the transition state for rotation, that is the activation energy for rotation, becomes small if the ground state energy is raised, whereas the transition state energy is held constant. Therefore it is necessary to lower the ground state energy and to raise the transition state energy to realize a high barrier to rotation. This is not an easy task to achieve, though conceptually it is understood.

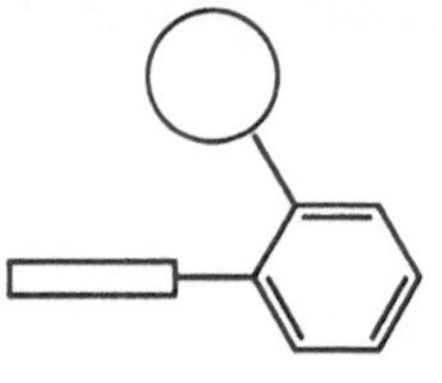

1-9 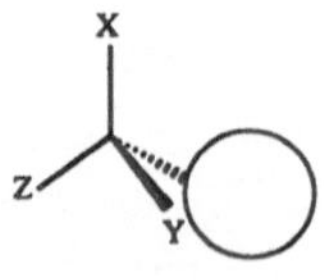   1-10

The relatively high barriers to rotation observed in 2,2,3,3-tetrahalobutane derivatives with respect to 2,3,3-trimethyl-2-butyl derivatives must be derived by both the effects in the transition state and the ground state for rotation. In the tetrahalo compounds, the ground state is relatively relaxed due to the small bulk of the halogens, while the transition state is destabilized by the electric effect as well as the steric, though the latter is smaller than in the case of the trimethyl-2-butyl derivatives.

It is known that, in order to make it possible to isolate rotational isomers at room temperature, the barrier to rotation must be 23.5 kcal/mol or more [26]. The barrier to rotation in butane, as well as in 2,3,3-trimethyl-2-butyl derivatives, is too low for isolation of rotational isomers at room temperature. It is therefore necessary to raise the barrier to make the isolation of rotational isomers possible at room temperature.

Various compounds were submitted to dynamic NMR study to estimate their barriers to rotation. Through these studies, various factors that influence the barrier heights have been clarified. The congestion at the ground state is shown to be effective in lowering the barrier to rotation in many cases. o-Di-*tert*-butylbenzene (**1-11**) [27] and 1,8-di-*tert*-butylnaphthalene (**1-12**) [28] are known to possess very low barriers to rotation of the *tert*-butyl groups, though their structure implies that the steric effect in the transition state for rotation is huge. It is true that the transition state energy for rotation is very high for these compounds but the ground state energy is also raised because of the congestion. The difference between the energies of these two states seems to be small.

1-11   1-12

It is known that, in order to make it possible to isolate rotational isomers about an sp^2–sp^3 bond, the substituent has to be large on one side and it has to be small on the other. Thus it is possible to isolate rotational isomers of di-*tert*-butyl(o-methylphenyl)carbinol (**1-13**) at room temperature [29]. In this compound, the two *tert*-butyl groups make the transition state for rotation very unstable, whereas the o-methyl group relaxes the ground state to some extent. Indeed, the sc-form is more stable than the *ap*.

ap-1-13   -sc-1-13

Of course, it is possible that the barrier to rotation is influenced by the solvent, especially when the polarity of the ground state and the transition state is different to a large extent. If an amide is taken as an example, this point is clear. In the ground state, it possesses a polar structure due to the contribution of the polar resonance structure. However, in the transition state the polar nature described by the resonance canonical structure is lost because of the geometry of the molecule. Thus a polar solvent should enhance the barrier to rotation by stabilizing the ground state. However, large solvent effects on the barrier to rotation have not been reported for amides [30], although there is such a case for push-pull ethylenes [31].

1.4.2 Triptycene Derivatives

Heaney and his coworkers studied the addition of benzyne to various compounds to show that electronegative benzynes could even be added to benzene rings. They reported that a product (**1-14**) of the addition of tetrafluorobenzyne to *tert*-butylbenzene showed three methyl proton peaks, of which intensities are the same in its ^{1}H NMR spectrum at room temperature [32]. Two of the three signals can be attributed to coupling with the fluorine nucleus that is located closely to the *tert*-butyl group. Therefore, the spectrum meant that there were two methyl groups whose magnetic environments are the same and which couple with a fluorine atom in proximity, and another methyl group which does not couple with fluorine atoms and that the exchange of their sites are slow on the NMR time scale.

1-14

When the temperature was raised the signals due to the methyl protons coalesced at 132 °C. Calculation of the barrier to rotation from the published data revealed that the free energy of activation for rotation was ca. 20 kcal/mol. This result was encouraging because the structure of the molecule could be modified and the raising of the barrier by 4 kcal/mol would make it possible to isolate the rotational isomers at room temperature.

A natural extension is a molecule which has substituent(s) on the etheno bridge of Heaney's compound **1-14**. Accordingly, the following compound (**1-15**) was prepared by a Diels-Alder reaction of 9-*tert*-butylanthracene with dimethyl acetylenedicarboxylate. The methyl signals of the *tert*-butyl group in ^{1}H NMR spectra were split into two, as expected, with intensities of 2:1. The signals did not coalesce though the spectra were recorded at 132 °C. From the temperature and the chemical shift difference of the two methyl protons, the free energy of activation for rotation was calculated to be more than 25 kcal/mol [33]. Of course, this compound cannot give rotational isomers no matter how high its rotational barriers may be. The *tert*-butyl group must be modified to make the symmetry of the substituent lower than C_{3n}.

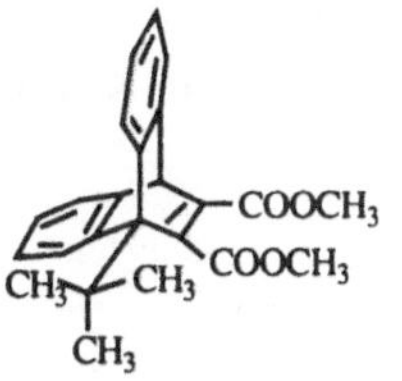

1-15 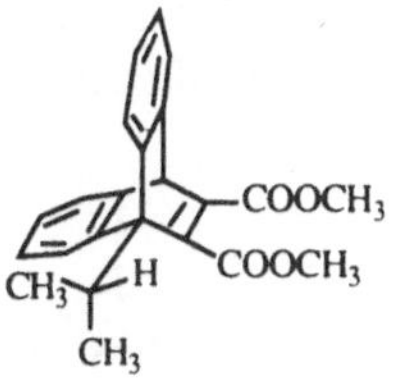   1-16

In order to achieve this, an isopropyl group was introduced instead of the *tert*-butyl group (**1-16**). However, the size of the isopropyl group was not large enough to raise the barrier to rotation to more than 24 kcal/mol: it was only 19 kcal/mol [33]. Therefore, a substituent which is as bulky as the *tert*-butyl group yet does not possess C_{3v} symmetry had to be chosen. As one such substituent, a 1,1-dimethyl-2-phenylethyl group, that carries a phenyl substituent for one of nine hydrogens in a *tert*-butyl group, was introduced.

The compound (**1-17**) indeed showed a barrier to rotation high enough for isolation of rotational isomers at room temperature. The rotational isomers, *ap* and *sc* forms, were isolated by chromatography and the barrier to isomerization was found to be 32 kcal/mol [34]. Hydrolysis of these compounds with potassium hydroxide afforded compounds in which only one of the methoxycarbonyl groups was hydrolyzed due to the steric effects. The monocarboxylic acid was converted to (−)-menthyl ester, which showed remarkably different solubility for diastereomeric esters. By recrystallization of the ester followed by hydrolysis, it was possible to isolate (+)- and (−)-forms of the monocarboxylic acid. The absolute stereochemistry is now determined and the − *sc* isomer of the dimethyl ester (− *sc*-**1-17**) is levorotatory at the Na-D line [35].

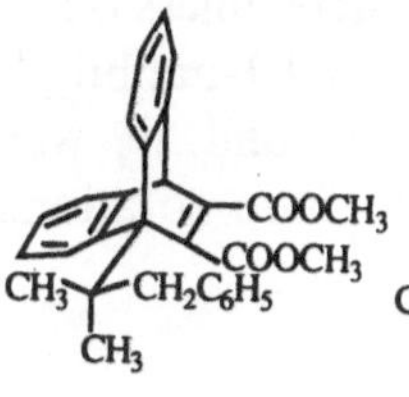

+*sc*-**1-17** 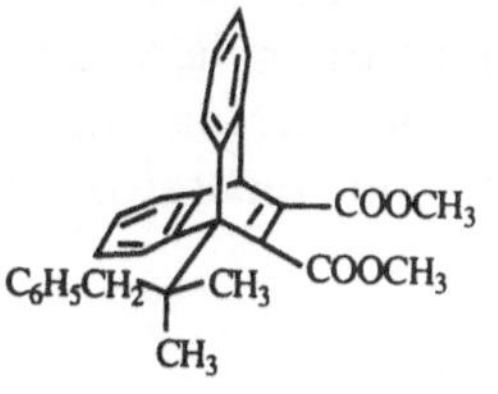   *ap*-**1-17** 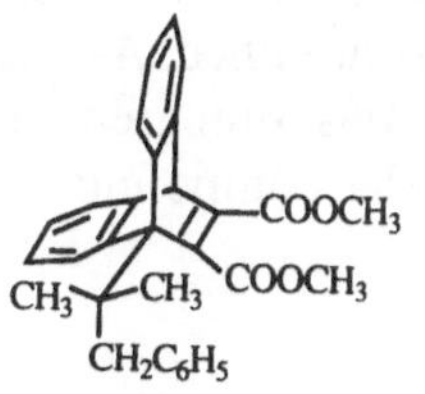   -*sc*-**1-17**

Following these works, the triptycene skeleton was introduced for the isolation of rotational isomers [36], because it was much easier to change the substituent, which could affect the rotational barriers, in the triptycene system.

The triptycenes can be prepared by addition of benzynes to substituted anthracenes, thus substituted anthracenes as well as substituted benzynes afford variously substituted triptycenes. The triptycene family (1-18) proved to be an excellent group of compounds which exhibit various molecular interactions as well as showing varieties of rotational barriers, ranging from 32 to 42 kcal/mol. The triptycene skeleton is ideal in manifesting the high barrier to rotation about the C_9-to-substituent bond, because its ground state is relatively relaxed irrespective of the fact that it is the combination of two *tert*-alkyl groups. The details will be discussed in Sect. 3.3.

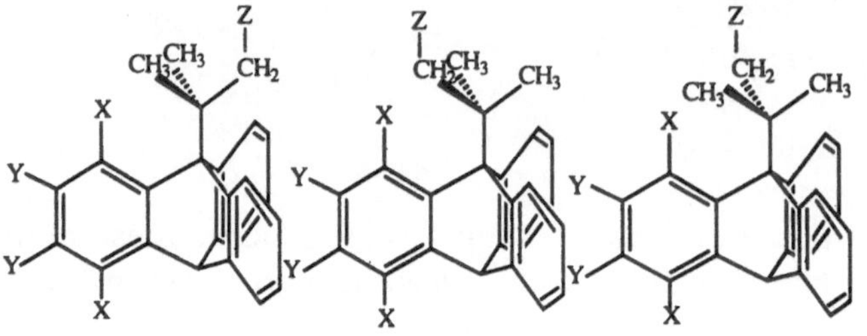

ap-1-18   +sc-1-18   -sc-1-18

1.4.3 9-Arylfluorenes

Three laboratories reported almost simultaneously that the rotational barrier about the C_9–C_{ar} bond in 9-(2,4,6-trimethyl-phenyl)fluorene (1-19) was higher than 25 kcal/mol by the dynamic NMR study of the compound and the related [37–39]: The signals due to two o-methyl protons of the 2,4,6-trimethylphenyl group are split in its ^{1}H NMR spectra and they did not coalesce even at 200 °C.

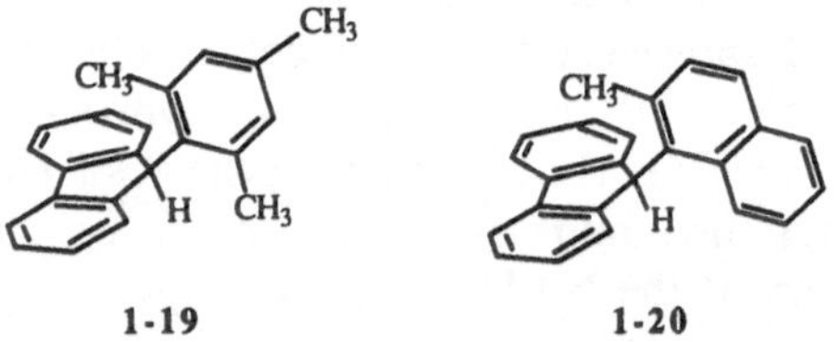

1-19   1-20

Here again, it is not possible to separate the rotational isomers unless one of the o-methyl groups is modified. As such an example, 9-(2-methyl-1-naphthyl)-fluorene (1-20) [37] was prepared and the stereoisomers, ap and sp, were partially separated. The compound was later separated into pure rotational isomers [40].

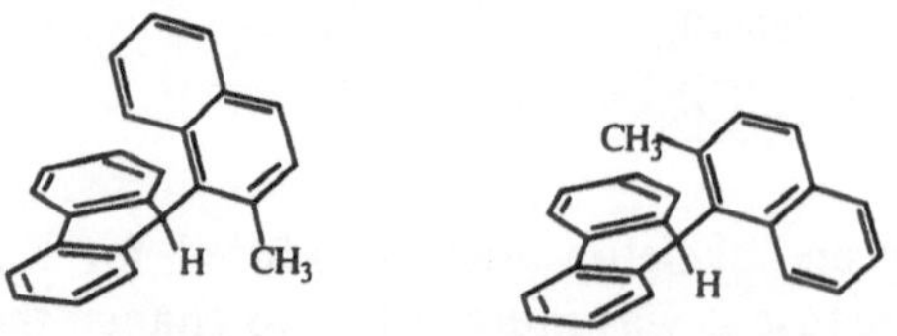

ap-1-20   sp-1-20

It was also possible to separate rotational isomers of this type, when one of the methyl groups in 9-(2,6-dimethylphenyl)fluorene was modified to a bromo-methyl group (**1-21**) [41]. Thus 9-arylfluorenes have become the second group of compounds which afford stable rotational isomers at room temperature.

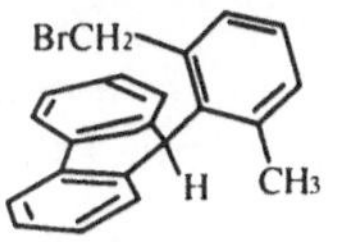

ap-**1-21** 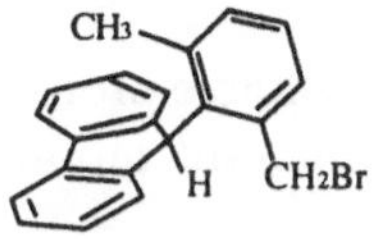   *sp*-**1-21**

Strictly speaking, there should be three pairs of such isomers, if one admits that a conformation about an sp^2-sp^3 bond is stable when the double bond is eclipsing as in propene and 1-butene. Indeed there is also ample evidence for the stability of such conformations in the aromatic compounds. Rotational barriers in isopropylbenzenes have been examined by NMR methods [42]. In the ground state, two methyl groups are magnetically equivalent to show that the C–H bond of the isopropyl group is eclipsing the benzene ring. The eclipsing form of the C–C bond of the isopropyl group with the benzene ring in isopropyl benzene was not detected. When this conformation is applied to the present case, it is predicted that the eclipsing form (*ap* or *sp* form) of the C–H bond of the 9-position of the fluorene ring with the 9-aryl group would be stable. Other eclipsing conformations (*ac* or *sc* forms) are unstable as well due to the steric effect. These conformations are depicted in the following scheme as projections along the C–C bond connecting the aryl group and the fluorene moiety.

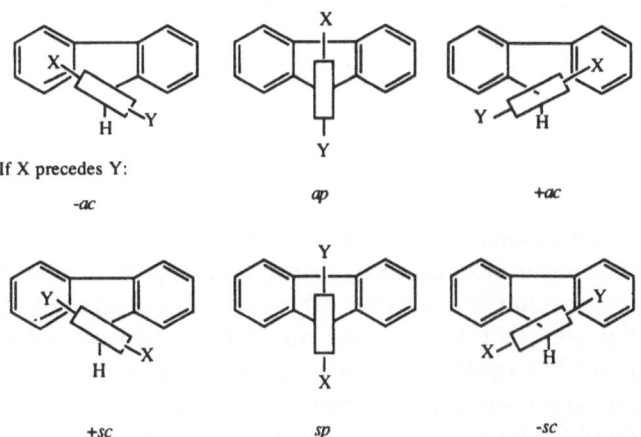

If X precedes Y:

   -ac   *ap*   *+ac*

   +sc   *sp*   *-sc*

It is now known, however, even in the *ap* and *sp* forms, there are two forms of each. The completely eclipsing form is avoided and the molecule librates within the *sp* or *ap* region. This motion is shown in the next scheme, by taking the *ap* form of the 9-arylfluorene. The barrier to the libration is of course so low that isolation of the librating isomers is not possible at room temperature but such a motion is detected if a compound carries a suitable probe which contains a pair of nuclei that can become diastereotopic when the motion becomes slow [43].

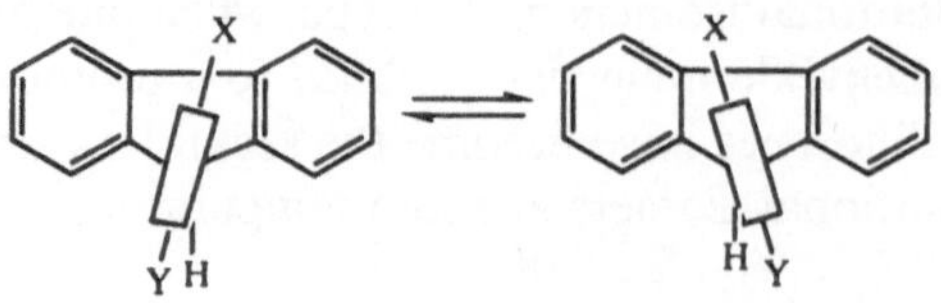

1.5 Concepts That Need Modification

If one admits that rotational isomers of organic compounds are diastereomers, then it should also be admitted that they behave differently in physical and chemical treatments. There are several points that deserve mention: some classical concepts have to be modified under these conditions.

1.5.1 Are There Dextrorotatory and Levorotatory Groups?

One of the points is the explanation of the property of *meso*-tartaric acid (**1-22**). In the classical concept, the *meso*-form is not optically active, because one of the halves of the molecule rotates the plane of the incident plane-polarized light to the right, whereas the other half rotates it to the left to the same degree as the dextrorotatory half does. This is so-called internal compensation.

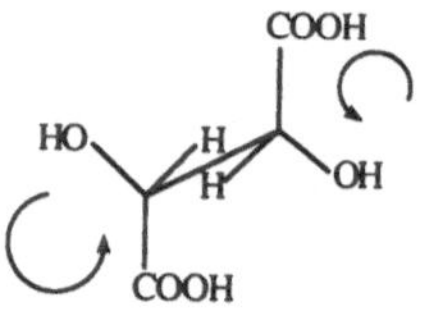

1-22

However, it becomes immediately obvious when one writes the stereostructure of the rotamers that *meso*-tartaric acid also possesses three stable rotational isomers, of which two are enantiomers with each other and another an optically inactive one due to the fact that it possesses an identical mirror image. Therefore, *meso*-tartaric acid is optically inactive because it possesses equal amounts of dextrorotatory and levorotatory isomers in addition to the optically inactive *ap*-conformation of the molecule. The isolation of rotational isomers is only a matter of temperature and someday the dextro- and levo-forms of *meso*-tartaric acid will be isolated as chemical entities. This point was mentioned by some chemists [44, 45].

It was recently pointed out that (*R,R*)-tartaric acid amides show different CD spectra from other derivatives of the same compound and the phenomenon was attributed to the difference in stable conformations of the molecules [46].

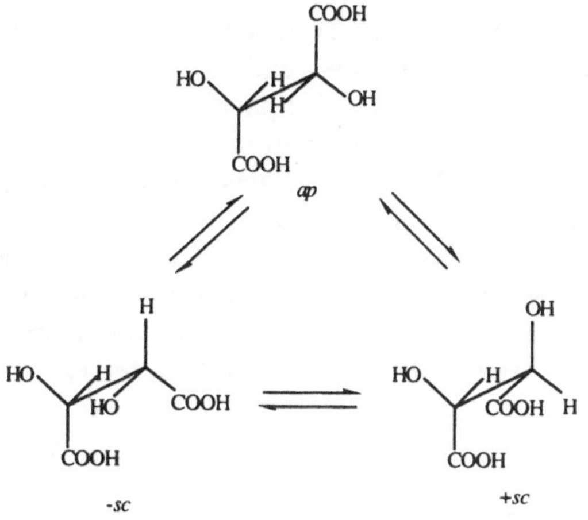

Indeed, N,N,N',N'-tetraalkyl-(R,R)-tartaramides were found to crystallize in $-sc$-conformations. meso-Tartaric acid could behave analogously under certain conditions. If one sticks to the concept of a dextrorotatory group and a levorotatory group, it would not be possible to admit these phenomena or predictions.

1.5.2 Reactivity of Rotational Isomers

Another point which needs special mention is the reactivity of conformational isomers. Organic chemists usually assume that a mixture of rotational isomers can be treated as a single compound, because they are not separable at room temperature. However, as already pointed out, rotational isomers are often diastereomers whose reactivity should in principle be different.

This point was singled out by Curtin [47] and is now known as the Curtin-Hammett principle [48]. The simplest case of application of the principle to the chemistry of rotational isomers may be explained by taking an example in which only two rotational isomers (A and B) exist.

$$Y \xleftarrow{\;k_A\;} A \underset{k_{-1}}{\overset{k_1}{\rightleftharpoons}} B \xrightarrow{\;k_B\;} X$$

The equilibrium constant (K or the population ratio of the rotamers) is expressed by $K = k_1/k_{-1}$. If the rotational isomer A reacts under certain conditions to give a product Y with a rate constant k_A and another isomer B a product X with a rate constant k_B, then the product ratio [X]/[Y], is expressed by the following equation.

$$[X]/[Y] = K \cdot k_B/k_A$$

In the old days, it was not possible to know the population ratios of rotational isomers except in special cases. Thus the product ratios were taken to mean the relative rate constants of a reaction for rotational isomers. For example, dehydrobromination of 2-bromoethane (**1-23**: R=CH₃) with ethanolic potassium hydroxide afforded a 6:1 mixture of *trans*- and *cis*-2-butene (**1-24**: R=CH₃) [49], whereas that of 1-bromo-1,2-diphenylethane (**1-23**: R=C₆H₅) showed a relative rate ratio of 130 for formation of *trans*- and *cis*-stilbenes (**1-24**: R=C₆H₅) [50]. These results are interpreted on the basis of interactions of the bulky substituents in these compounds in the transition state of the reaction: it will be understood from the following scheme that the transition state for the formation of the *cis*-isomer is more energy-demanding than that for the formation of the *trans*-isomer because of the interactions between two R groups in the former. When the R group is bulky, the effect is large.

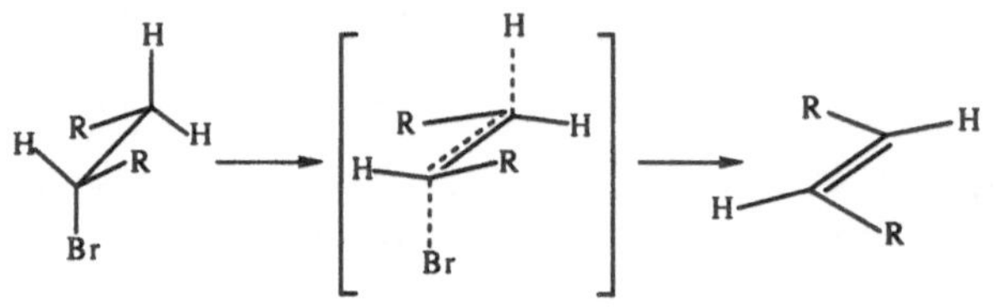

+sc-**1-23**   *trans*-**1-24**

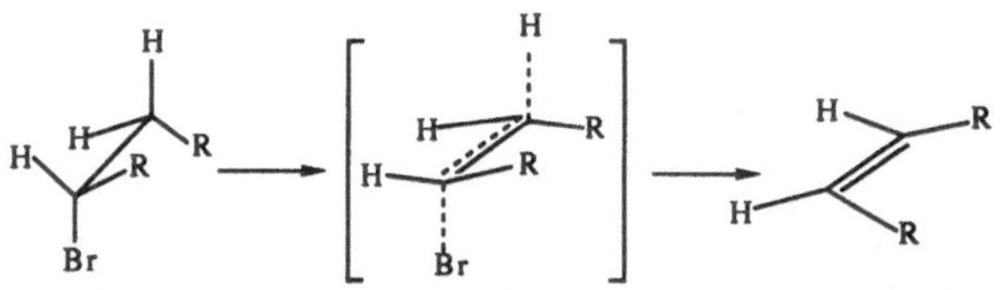

-sc-**1-23**   *cis*-**1-24**

If we had known the population ratios of the rotational isomers, it would have been possible to mention more precisely the relative reactivities of these rotamers. Since it is now possible to ascertain fairly accurate population ratios by spectroscopic methods including nuclear magnetic resonance, it is possible to determine the relative rates of the rotamers as well by this principle. However, there is one point that one has to bear in mind.

This principle would enable one to predict the product ratio, if the population ratios of the rotamers concerned and the rate constants of the reaction of the rotamers were known. However, it is not possible to find out the rate constant of the rotamers except in some special cases, where one isomer is predominant under the conditions of the reaction. Therefore, the Curtin-Hammett principle is very useful in explaining the product ratios of reactions but is rarely used in predicting the product ratio.

As can be seen from the equation of the Curtin-Hammett principle, it is necessary to have adequate rate constants for the isomerizations for the principle to hold, because otherwise the concentration of one isomer may be different from that of the equilibrium due to consumption of the isomer by reactions. This point was treated mathematically and it is now known that if the rate constants for the equilibrium of the rotamers are larger than the rates of the reactions of the rotamers by 10 times, the Curtin-Hammett rule holds in practice [51].

In the actual cases of rotational isomers, three forms are present and in extreme cases these three would give different products under the same reaction conditions. This can be shown by the following scheme and is a general case of the Curtin–Hammett principle applied to the chemistry of rotational isomers.

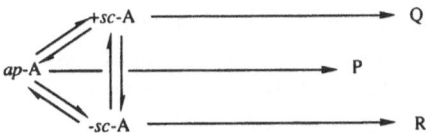

However, in more or less general cases, a reaction of a mixture of rotational isomers will give a mixture of rotational isomers of a product. This case can be depicted in the following scheme. Here it is possible that the reaction rates are greatly different from one to another, where a compound reacts practically in one conformation for which reaction rates are high.

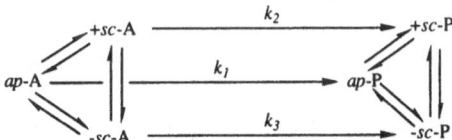

These considerations clearly indicate that the knowledge of the reactivity of rotational isomers is indispensable for understanding and improving selectivities of a reaction, if one has to treat a compound as a mixture of rotational isomers. However, this type of knowledge had not yet been obtained. The best approximation was to use model compounds such as 4-substituted *tert*-butylcyclohexanes [52] which give one of the conformational isomer models. In the last two decades, it has become possible to isolate rotational isomers in some cases and to obtain information about the reactivity of the rotational isomers.

The primary objective of this book is to present differences in reactivities of such isomers. As one would suppose, the difference is very small in some cases. But in other cases, the rate constants of rotational isomers can be different by 1000 times or more and rotational isomers can even react by different mechanisms, though the product is a single compound if we neglect rotational isomers. In still other cases, rotational isomers give different products.

The reader of this book should be now convinced that a mixture of rotational isomers is a real mixture and not a pure compound. Unfortunately, this concept has not yet received wide acceptance.

For example, an IUPAC compendium of terminology [53] states the following for "chemical species". "A set of chemically identical atomic or molecular structural units in a solid array or of chemically identical molecular entities that can explore the same set of molecular energy levels on the time scale of the experiment. For example, two conformational isomers may interconvert sufficiently slowly to be detectable by separate NMR spectra and hence to be considered to be separate chemical species on a time scale governed by the radiofrequency of the spectrometer used. On the other hand, in a slow chemical reaction the same mixture of conformers may behave as a single chemical species, i.e. there is a virtually complete equilibrium population of the total set of molecular energy levels belonging to the two conformers."

This statement can be misleading on two points. One is that the statement does not mention temperature. It should be recognized that any rotational isomer can isomerize sufficiently slowly to jeopardize the statement at a specific temperature. One has to admit that a single chemical species at one temperature can be two or more species at a different temperature. Secondly, this statement does not explain one of the cases which is mentioned by the Curtin-Hammett principle. That is, a rotational isomer can give a different product from that produced by another rotational isomer. The case which is mentioned in the term "chemical species" is thus a rather special one. It is much easier to recognize that a mixture of rotational isomers is not a pure compound.

It is important that one recognizes the presence of rotational isomers in a compound which was thought to be pure. If one wishes to increase the selectivity of a reaction of a given organic compound, the reactivity of the rotational isomers has to be taken into consideration. It is especially true when one is running a cyclization reaction or an asymmetric synthesis, because these reactions are typically ones in which different rotational isomers give different products.

1.6 References

1. van't Hoff JH (1874) Arch Neerl Sci Exact 9: 445; (1875) Bull soc chim France 23: 295
2. Le Bel JA (1874) Bull soc chim France 22: 337
3. van't Hoff JH (1875) La chimie dans l'espace, Bazendijk, Rotterdam
4. Mizushima S, Morino Y, Higashi K (1934) Sci Pap Inst Phys Chem Res 25: 159
5. Kemp JD, Pitzer KS (1936) J Chem Phys 4: 749. (1937) J Am Chem Soc 59: 276
6. Dauben W, Pitzer KS (1956) Conformational analysis. In: Newman MS (ed) Steric effects in organic chemistry. Wiley, New York, p 1
7. Evans MW (1974) Spectrochim Acta 30A: 79
8. Eliel EL (1976) Isr J Chem 15: 7
9. Lide DR Jr, Christensen D (1961) J Chem Phys 35: 1374
10. Whangbo M-H, Schlegel HB, Wolfe S (1977) J Am Chem Soc 99: 1296
11. Kondo S, Hirota E, Morino Y (1974) J Mol Spectrosc 28: 471
12. Christie GH, Kenner J (1922) J Chem Soc 121: 614
13. Kuhn R (1933) Molekulare Asymmetrie. In: Freudenberg H (ed) Stereochemie. Franz Deutike, Leipzig, p 803
14. Mizushima S (1954) Structure of molecules and internal rotation, Academic, New York
15. Jensen FR, Bushweller CH (1966) J Am Chem Soc 88: 4279
16. Hofner D, Lesko SA, Binsch G (1978) Org Magn Reson 11: 179
17. Tanigaki K, Yagi M, Higuchi J (1989) J Magn Reson 84: 282
18. Lister DG, Macdonald JN, Owen NL (1978) Internal rotation and inversion, Academic, London

19. a) Kessler H (1970) Angew Chem 82: 237 Angew Chem Int Ed Engl 9: 219. b) Binsch G (1975) Band shape analysis. In Jackman LM, Cotton FA (eds), Dynamic nuclear magnetic resonance spectroscopy, Academic, New York, p 45
20. Pitzer KS (1940) Chem Rev 27: 39
21. Hoogasian S, Bushweller CH, Anderson WG, Kingsley G (1976) J Phys Chem 80: 643. Anderson JE, Doeck CW, Pearson H, Rawson DI (1978) J Chem Soc, Perkin II, 974
22. Brownstein S, Dunogues J, Linsay D, Ingold KU (1977) J Am Chem Soc 99: 2073
23. Flamm-ter Meer MA, Beckhaus H-D, Peters K, von Schnering H-G, Fritz H, Rüchardt C (1986) Chem Ber 119: 1492
24. Adams R, Yuan HC (1933) Chem Rev 12: 261
25. Hawkins BL, Bremser W, Borsic S, Roberts JD (1971) J Am Chem Soc 93: 4472
26. a) ref 19a. b) Ōki M (1983) Top Stereochem 14: 1
27. Gibbons WA, Gil VMS (1965) Mol Phys 9: 163
28. Anderson JE, Franck RW, Mandella WL (1972) J Am Chem Soc 94: 1880
29. Lomas JS, Dubois JE (1976) J Org Chem 41: 3033
30. Stewart WE, Siddall TH, III (1970) Chem Rev 70: 517
31. Sandström J (1983) Top Stereochem 14: 83
32. Brewer JPN, Eckhart IF, Heaney H, Marples BA (1968) J Chem Soc C 664
33. Ōki M, Suda M (1971) Bull Chem Soc Jpn 44: 1876
34. Yamamoto G, Nakamura M, Ōki M (1975) Bull Chem Soc Jpn 48: 2592
35. Toyota S, Yamane H, Ōki M (1993) Bull Chem Soc Jpn 66: in press
36. Yamamoto G, Ōki M (1975) Bull Chem Soc Jpn 48: 3686
37. Siddall TH III, Stewart WE (1969) J Org Chem 34: 233
38. Chandross EA, Sheley CF Jr (1968) J Am Chem Soc 90: 4345
39. Rieker A, Kessler H (1969) Tetrahedron Lett 1227
40. Nakamura M, Nakamura N, Ōki M (1977) Bull Chem Soc Jpn 50: 1097
41. Nakamura M, Ōki M (1974) Tetrahedron Lett 505. Murata S, Kanno S, Tanabe Y, Nakamura M, Ōki M. (1982) Bull Chem Soc Jpn 55: 1522
42. Mannschreck A, Ernst L (1971) Chem Ber 104: 228
43. Murata S, Mori T, Ōki M (1984) Bull Chem Soc Jpn 57: 1970
44. Urushibara Y (1947) Lecture Note. University of Tokyo. The main work described in this book was inspired by this lecture.
45. Mislow K (1950) Science 112: 26
46. Gawronski J, Gawronska K, Rychlewska U (1989) Tetrahedron Lett 30: 6071
47. Curtin DY (1954) Rec Chem Progr 15: 111
48. Hammett LP (1970) Physical organic chemistry, 2nd edn, McGraw-Hill, New York, p 119
49. Lucas HJ, Simpson TP, Carter JM (1925) J Am Chem Soc 47: 1462
50. Curtin DY, Kellom DB (1953) J Am Chem Soc 75: 6011
51. Seeman JI (1983) Chem Rev 83: 83
52. Winstein S, Holness NJ (1955) J Am Chem Soc 77: 5562; Eliel EL, Lukach CA (1957) J Am Chem Soc 79: 5986
53. Gold V, Loening KL, McNaught AD, Sehmi P (eds) (1987) Compendium of chemical terminology (IUPAC Recommendations), Blackwell, Oxford

2 Rotamer Populations

2.1 Estimation of Rotamer Populations

The populations of rotamers can be estimated by various methods, usually with the use of a spectroscopic technique. For the spectroscopic technique, it is necessary to assign a given signal to a specific rotamer. This assignment and the estimation of the population with the use of the signal intensity will be discussed in this section.

2.1.1 Vibration Spectroscopy

There are two methods in this category: the use of Raman spectra and infrared spectra. Due to the time scale of these spectroscopies, it is easy to recognize rotational isomers of a molecule even at room temperature except in some cases discussed below. It is a great problem, however, for these spectroscopies to correlate intensities of the spectral lines with the population of rotamers, because it is the change in dipole moment by the vibration that affects the intensity of the infrared absorptions and is the change in polarizability by vibration in Raman spectral lines. Since these can vary from one rotamer of a molecule to another rotamer, the spectral intensity does not necessarily reflect the populations of rotamers in these spectroscopies.

The problem in recognizing the rotational isomers by these spectroscopies is the fact that more than one spectral line is observed in cases other than the presence of rotamers. Couplings of vibrations are easily pointed out when one cautiously examines the molecular structure, but the more serious problem lies in the case of Fermi resonance. Most complex organic compounds are able to give Fermi resonance lines and thus it is necessary to exclude the possibility of Fermi resonance in order to claim that one has observed rotational isomers.

Exclusion of the possibility is most precisely carried out by shifting one of the possible lines by substituting an isotope for an original atom, of which a mode of vibration is suspected of giving rise to a Fermi resonance line, and observing the disappearance of one of two lines. However, this is not very easy in practice and is sometimes even impossible. The next choice is the observation of the spectra at different temperatures, because, if the presence of two lines are due to the presence of rotamers, the intensities of the two lines should be different, i.e. the intensity of one band increases at the expense of another during the change in temperature. Unfortunately, temperature variation in the vibration spectroscopies are not so easy as in nuclear magnetic resonance. The third choice, which is not completely sure, strictly speaking, is the observation of solvent effects. If a band increases in its intensity in a polar solvent at the expense of another, then the former is assigned to a more polar rotational isomer than the latter.

Assignment of the lines in these spectra is usually made by calculation of normal vibrations. But it is not easy to do so for the complex molecules in which organic chemists are interested, because of the approximations involved. In these cases, model compounds, which give a similar result to a given bond, are often used. One of the most well known examples is the case of axial and equatorial chlorocyclohexane interconversion. These conformational isomers could be modeled by more complex compounds such as steroids, in which the cyclohexane inversion is locked. It is common practice today to use a *tert*-butyl group for locking the inversion of the cyclohexane ring.

axial   **2-1**   equatorial

2-2   **2-3**

Chlorocyclohexane (**2-1**) is known to exhibit two C–Cl stretching absorption bands in infrared spectra at 731 and 683 cm^{-1} [1]. For the assignment of the absorption bands, *cis*-4-*tert*-butyl-1-chlorocyclohexane (**2-2**) is a model for the axial chlorocyclohexane and *trans*-4-*tert*-butyl-1-chlorocyclohexane (**2-3**) is a model for the equatorial form. Since the former shows an absorption band due to C–Cl stretching at 690 cm^{-1}, the absorption at 683 cm^{-1} of chlorocyclohexane is assigned to the axial form. Similarly *trans*-4-*tert*-butyl-1-chlorocyclohexane absorbs at 725 cm^{-1} which is also close to the absorption band of chlorocyclohexane at 731 cm^{-1} [2].

After establishing the assignment of spectral bands to specific rotational isomers, problems still remain in conformational analysis by vibration spectra. That is, the spectral intensities do not correspond directly to the population of the rotamers. The actual situation will be explained by taking again the case of chlorocyclohexane as an example. According to the literature [3], the intensity of the C–Cl stretching absorption band due to *cis*-4-*tert*-butyl-1-chlorocyclohexane is 75.0, if expressed by the unit of molecular extinction coefficient, whereas that of the *trans*-form 119.6. The integrated intensity of the C–Cl stretching band of the *trans*-form is also stronger than that of the *cis*-form by 47%. Thus the apparent intensities of the two absorption bands of chlorocyclohexane cannot be taken as that they show the populations of equatorial and axial isomers.

In the days when NMR spectroscopy was not available, this problem was treated in the following way. The free energy difference ΔG and the equilibrium constant K can be correlated by

$$\Delta G = -2.303RT\log K$$

where R is the gas constant and T the temperature. Since the observed value is the ratio of intensities of two absorptions due to rotamers and the intensity is different from each other, the ratio is correlated with the equilibrium constant by the following equation.

$$K = AK'$$

where A is the correlating factor and K' the intensity ratio observed.

Then the first equation is rewritten as follows using C for $1/A$.

$$\Delta G = -2.303RT\log CK = -2.303RT(\log C + \log K)$$

Since the free energy difference is correlated with enthalpy and entropy difference by the following equation

$$\Delta G = \Delta H - T\Delta S$$

the slope obtained by plotting the $\log K$ values at various temperatures against $1/T$ should give the enthalpy difference. Though it is not possible to obtain the entropy difference in this way, because the intercept contains the term $\log C$, the enthalpy data were useful because entropy changes in the rotamer distribution are usually close to zero.

This method possesses apparent shortcomings. One is that temperature variation is not easy in infrared spectroscopy and the other is that entropy differences are not zero, though usually very small, in the equilibrium of rotamers, especially in the cases of polar compounds in polar solvents. Thus the method became unpopular after NMR spectroscopy was introduced into organic chemistry.

2.1.2 Nuclear Magnetic Resonance Spectroscopy

NMR spectroscopy has an advantage over the vibration spectroscopies in the study of rotational isomers due to the fact that the intensities of signals are proportional to the number of nuclei at the magnetic site. Thus it is possible to estimate populations of rotational isomers by this technique with a precision of up to ca. 5% with the classical spectrometers and up to ca. 2% with the contemporary instruments equipped with superconducting magnets, unless the intensity of a signal due to one isomer is particularly reduced due to relaxation.

The weak point of this technique is the time scale. Because the temperature of coalescence/decoalescence at a given barrier is a function of the chemical shift differences (and coupling constants) of two signals concerned, the coalescence temperature becomes higher when a high field magnet is used instead of a low field magnet. This enables one to study a barrier to rotation as low as ca. 5 kcal/mol with the use of a 400 MHz machine by observing line shape changes.

If the barrier to rotation is lower than 5 kcal/mol, the most popular way of determining it is the use of a method such as far-infrared or microwave spectroscopy [4]. However, it is also possible to determine the barrier by the measurement of T_1 (spin-lattice relaxation time) in the solid state by NMR spectroscopy [5].

In Table 2.1, an estimate of barriers to rotation and the coalescence temperature is shown for the convenience of the reader. Because of the fact that the coalescence temperature is dependent on the chemical shift difference of the signals and in part on the coupling constant, this should not be taken as a precise prediction but a rough estimate.

The first experiment that should be carried out is to identify which of the signals are due to rotational isomers. This is conveniently done by changing temperatures. When the coalescence is observed for two signals, they are usually the pair which are due to rotational isomers. In a case when a molecule gives a set of signals of more than two, these can be treated similarly.

The assignment of the observed signals to respective rotamers is made by various techniques: all parameters which differ in the three dimensional structures may be used. They are the chemical shifts, signal patterns, long-range couplings and nuclear Overhauser effects. The representative cases are discussed below.

A typical example of using the chemical shift difference is found in 9-arylfluorene derivatives. Due to the magnetic anisotropy effect of the benzene rings, the methyl proton signal of *ap*-9-(2-methyl-1-naphthyl)fluorene (**2-4**) gives an NMR signal at a higher magnetic field than that in the corresponding *sp*-form [6]. This chemical shift difference owes to the ring current effect of the

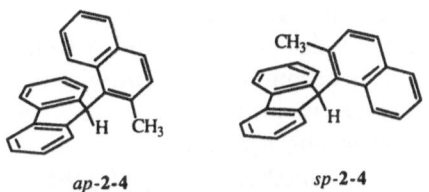

ap-**2-4**   *sp*-**2-4**

Table 2.1. Rough correlation between coalescence temperature in ^{1}H NMR spectra and the activation energy for rotation

Coalescence Temperature (°C)	$\Delta G^{\neq}$ (kcal/mol)
200	> 25
100	20
20	15
− 50	10
− 100	7

benzene nuclei that constitute the fluorene ring. The methyl group which is attached to the naphthalene nucleus is above the fluorene ring in the *sp* form and should give a signal at a higher field than that in the *ap* form.

An example of spectral patterns was provided by Nair and Roberts [7]. At temperatures lower than the coalescence, 1,1,2-tribromo-1-chloro-2,2-difluoroethane (**2-5**) shows one singlet and an AB quartet in its ^{19}F NMR spectrum. The reason for this spectral pattern is obvious when one looks at the following scheme which shows three rotational isomers using the Newman projection.

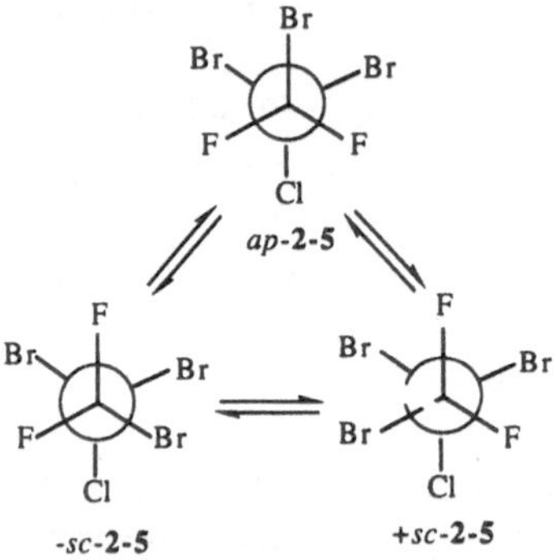

In the *ap*-form, two fluorine atoms are equivalent and thus should exhibit an A_2 type line. However, in the *sc*-forms, two fluorine atoms are not equivalent and couple with each other, thus showing an AB quartet line. This type of technique can be used very widely and examples may be seen in the following texts in this book.

If one of the rotational isomers does have a symmetry plane and others do not, the assignment of conformations can be readily made as above. However, if one of the ends of the ethane is a chiral center, it is not possible to do so. In this case, every pair of otherwise equivalent nuclei becomes diastereotopic, thus showing a unisochronous property. In these cases, various magnetic properties of the nuclei in rotational isomers must be taken into account.

A set of three dimethyl 2,3-dichloro-9-(1,1-dimethyl-2-phenylethyl)-9,10-dihydro-9,10-ethenoanthracene-11,12-dicarboxylate (**2-6**) rotamers will serve as one of the examples of such cases. The C_9 of the dihydroethenoanthracene skeleton is a chiral center in this compound. All three rotational isomers are isolated [8]. They all show AB quartet patterns for the methylene protons in the 9-substituent. In this case, the assignment was made by the aromatic proton signals that appear at the lowest magnetic field, δ 7.7–8.1. The features of the signals are the shift of a singlet signal and a multiplet signal, both of which are assigned to the *peri*-protons, because the van der Waals' shift is expected for these protons. From studies on model compounds, the *peri*-proton flanked by a methyl and a benzyl group in these compounds gives a multiplet signal at a lower magnetic field than that flanked by two methyl groups. The *peri*-proton which has a chloro-substituent should show a singlet, whereas the *peri*-proton which has no chlorine *ortho* to it should show a multiplet. Thus the *peri*-protons in *sc**(*S**)-**2-6** should show a singlet signal at higher magnetic field than a multiplet. Similarly *sc**(*R**)-**2-6** should show a singlet at a low field and a multiplet at a high field, whereas *ap*-**2-6** a singlet and a multiplet at about the same field. The experimental results were just as expected and the stereochemistry of the three rotamers was assigned.

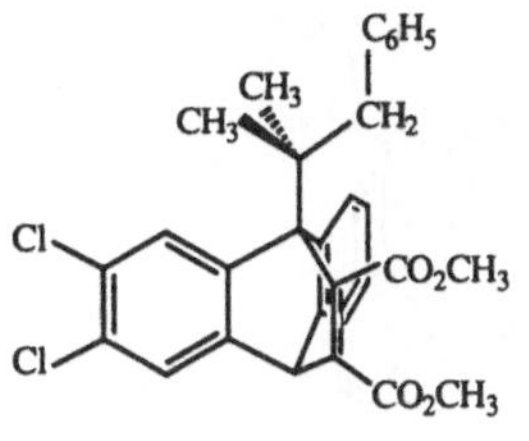

sc*(*S**)-**2-6**

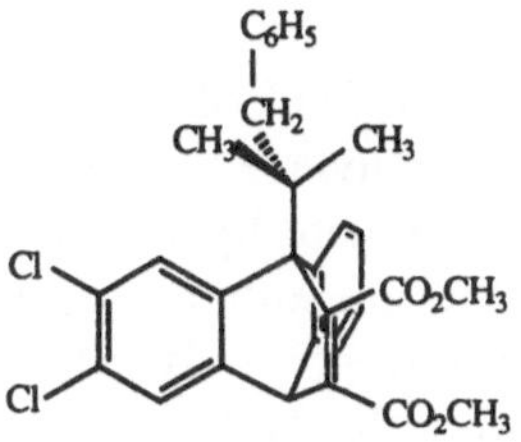

ap-**2-6**

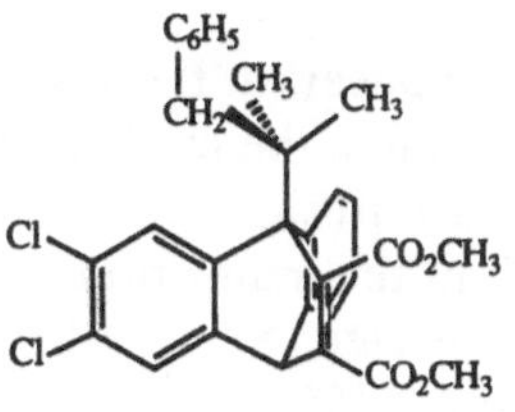

sc*(*R**)-**2-6**

Long range coupling may be used if such nuclei are present, though examples of which are not found in literature. The basic idea is that if an NMR active nucleus couples with another through space, it means they are close to each other. An example can be seen in 1,2,3,4-tetrafluoro-9-(1,1-dimethyl-2-phenylethyl)triptycene (**2-7**) rotamers [9], though they can be assigned by the spectral patterns mentioned above. The *ap*-isomer shows a methyl-proton signal coupling with the 1-fluoro substituent which means that both methyl groups are close to the 1-substituent. By contrast, the *sc*-isomer shows two methyl signals, one of which couples with the 1-fluoro substituent and the other is a singlet, which means that one of the methyl groups is close to the 1-substituent but the other is not.

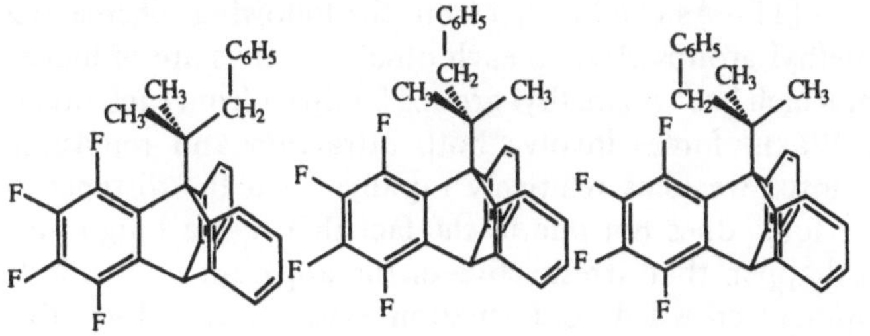

ap-**2-7**

-sc-**2-7**

+sc-**2-7**

Two of the possible three rotamers of 1,8-difluoro-9-(3-hydroxy-1-methylpropyl)triptycene (**2-8**) are isolated [10]. Although assignment of the structures was barely possible by considering the long-range coupling of the fluorine atoms (the fluorine atoms give almost the same coupling constant with the vicinal hydrogen in the 9-substituent), it is much easier and reliable to use the NOE technique.

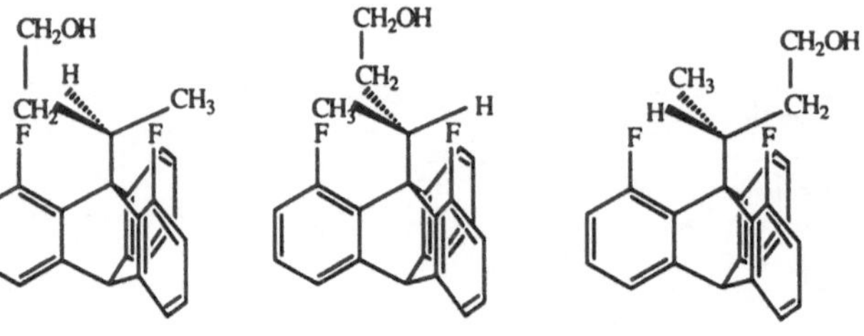

ap-**2-8**   sc*(R*)-**2-8**   sc*(S*)-**2-8**

When the *peri*-proton, which appears at the lowest magnetic field, of a major isomer in the synthesis was saturated, the methyl and the methylene (at the 1-position in the 2-hydroxyethyl group) signals showed enhancement by 4% and 11%, respectively, whereas that in the minor product showed the enhancement in the methine and the methylene signals by 12% and 11% respectively. Thus the major isomer is assigned to sc* (S*)-**2-8** and the minor sc*(R*)-**2-8**.

2.2 Factors That Affect Rotamer Populations

The first and the most frequently encountered factor that affects the population of rotamers is repulsion. This repulsion is mainly caused by the van der Waals' repulsive force. Thus, usually two bulky substituents which are connected to different carbon atoms of ethane have a tendency to move apart as far as possible. However, dipolar attraction can be an important factor that determines the population ratio of rotamers, if two dipoles can come close within the molecule.

The preference of the *ap*-isomer over the *sc* in butane (**2-9**) can be taken as one of the typical examples which show van der Waals' repulsion as a key factor to control the population [11]. As can be seen from the following scheme, the ± *sc* forms bear two methyl groups close to each other and thus are of higher energy than the *ap* form which has two methyl groups far apart from each other.

However, van der Waals' forces involve both attraction and repulsion. Whereas the repulsive force weakens relatively rapidly when the distance is increased, the attractive force does not due to the fact that it is a long-range force. Therefore, it can happen that, irrespective of the apparent unfavorable steric situation, a seemingly crowded conformation exists to an observable degree. Existence of a conformer that possesses *tert*-butyl groups on the same

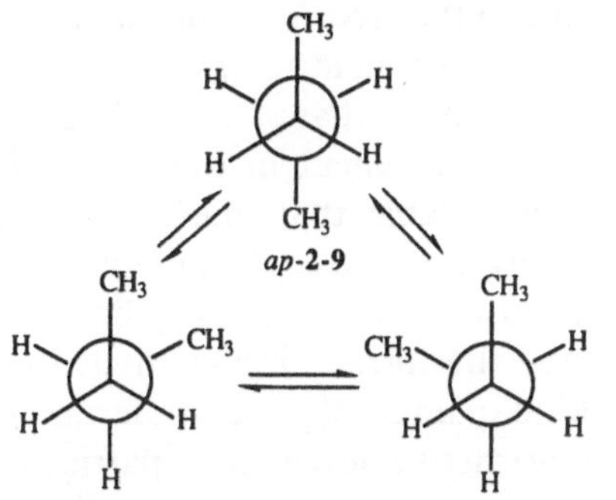

ap-2-9

+sc-2-9   -sc-2-9

side of the benzene ring in 1,3,5-trineopentylbenzene derivatives (**2-10**) is an example [12].

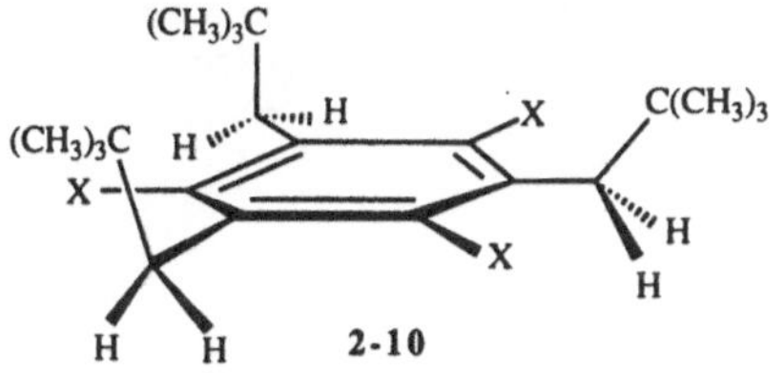

2-10

Recently, it has been postulated that halogen–halogen interactions are rather attractive. *cis*-1,2-Dichloroethylene was known to be more stable than the *trans*-isomer [13] and the cause was attributed to the attractive force of the van der Waals type. However, 1,2-dichloroethane is another story and is known to exist as an almost 1:1 mixture of two kinds of rotational isomers, the $\pm sc$ pair and the *ap* form, in liquid and the *ap*-form is more stable than the *sc* by ca. 1 kcal/mol in the gas phase [14]. If one considers the dipole-dipole interactions in the *sc*-form, it can be understood by assuming that the *sc*-form is stabilized by solvation in the liquid phase.

A typical example of dipole-dipole interactions is seen in 2-halocyclohexanone [15]. In these compounds, if the halogen takes an axial orientation, 1,3-interaction destabilizes the form. On the other hand, if the molecule takes the halogen-equatorial form, there are two dipoles, C=O and C–Cl, in nearly parallel arrangement. This is destabilized because of the Coulombic repulsive force. Therefore, 2-bromocyclohexanone should exist in an equilibrium that is a balance between these two opposing factors. It is known that the compounds exists as a mixture of ca. 3:1 Br-axial (**2-11**) and Br-equatorial (**2-12**) forms in tetrachloromethane.

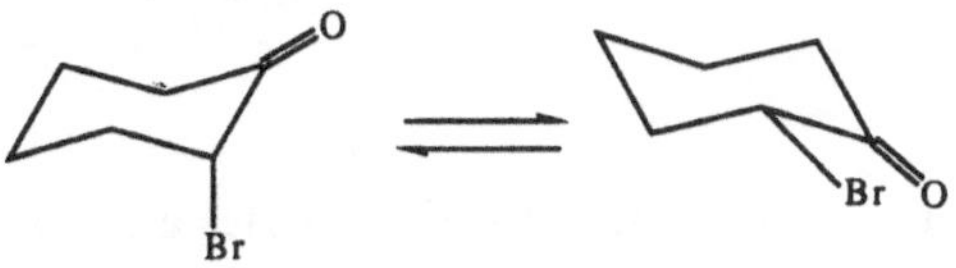

2-11   2-12

Irrespective of the unfavorable situations due to the factors discussed above, certain molecules can exist in such conformations. Then an attractive force must be considered. A typical example is seen in 2-methoxy-1-ethanol (**2-13**). Repulsive forces due to the steric effects and dipole arrangement in this molecule predict that the *sc*-conformation should be less stable than the *ap*-isomer. However, the actual molecule is almost exclusively in the *sc* conformation [16]. This is due to the presence of a hydrogen bond between the hydroxyl group and the methoxyl in the *sc* conformations, whereas the intramolecular hydrogen bond is not possible in the *ap*. Therefore, this kind of unusual population ratio is a reason for assuming the presence of an attractive force in a particular molecular conformation. There are many examples where the *sc* conformation is favored over the *ap*, in spite of the larger repulsive force that is expected in the former than in the latter.

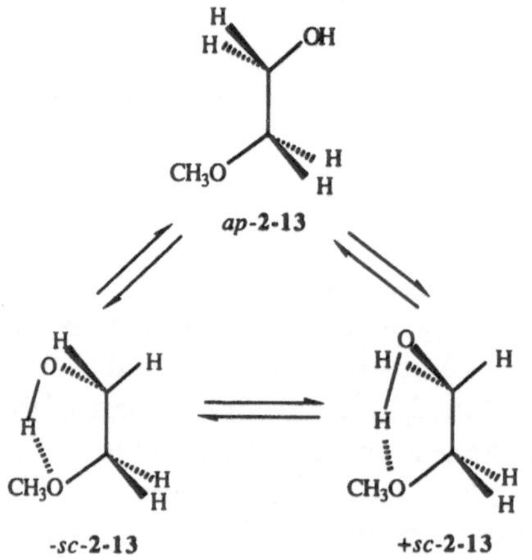

ap-**2-13**

-sc-**2-13**   +sc-**2-13**

Another example which shows anomalous population ratios of conformers is that found in oxane derivatives which carry an electronegative group at the 2-position. The NMR spectra of this type of compound were examined because of the interest in sugar chemistry. Sugars possess this moiety as its basic skeleton. On steric grounds, the axial form of 2-methoxyoxane (**2-14**: X=OCH$_3$) should be unfavorable relative to the equatorial form because of the interactions known as 1,3-diaxial repulsions in cyclohexane chemistry. However, the actual case is the reverse of this expectation: the axial form is favored [17]. This is called the anomeric effect [18].

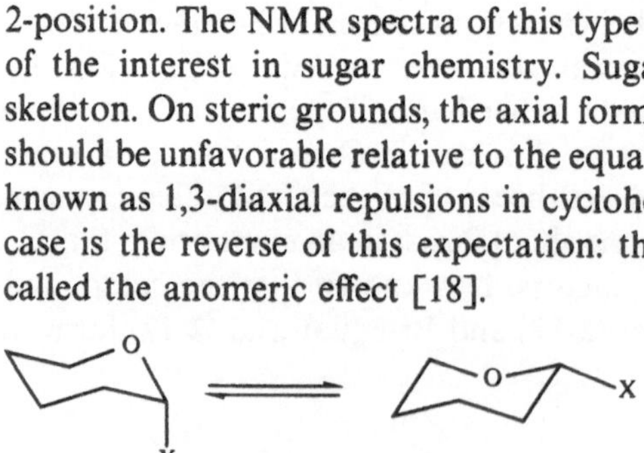

2-14

The anomeric effect is observed when a six-membered ring carries a heteroatom with at least a lone pair of electrons and an electronegative substituent at the 2-position. Today, it is believed that the stabilization is due to the interaction of the lone pair orbital and the σ^* orbital of the C–X bond, when they are arranged to make 180° [19]. This arrangement is favorable for the charge transfer interactions in which the occupied n-orbital of the heteroatom and the unoccupied σ^* orbital of the C–X bond participate.

Other factors such as quadrupolar interactions may affect the population ratios of rotamers. However, these are usually minor and can be neglected in explaining the population ratios.

2.3 Information on Molecular Interactions Obtained from Rotamer Populations

If the population ratios are abnormal from the stand point of the factors mentioned in Sect. 2.2, then the existence of an attractive force may be considered operative. This force is very often a charge-transfer type interaction.

A classical example of this sort is an interaction between amino-nitrogen and carbonyl functions. In cyclic amino-ketones, a conformation in which the two moieties are close to each other does exist because of the charge-transfer interaction [20], although the repulsive force disfavors such a conformation. However, if the electron donor is a sulfur [21] or an oxygen [22] instead of an amino nitrogen the interaction becomes very weak. Thus existence of a conformation favorable for the interaction is doubtful from the stand point of infrared spectroscopy, though X-ray crystal analysis provides information that such an interaction does exist [23].

The above statement includes an important piece of information when one considers the molecular interaction from the experimental results. Namely, the observation of attractive interaction by molecular association or by conformational population is possible when the attractive force is stronger than the repulsive force. If a repulsive force is strong, it may not be possible to observe an association of molecules or a preference for a conformation, no matter how much the interaction is favorable. Decrease in entropy by the fixing of the molecular conformation is another factor that should not be overlooked.

Hydrogen bond formation is known to be the summation of various forces including charge-transfer, dipole-dipole interactions, polarization, and van der Waals repulsion [24]. If repulsive force overwhelms the attractive force the presence of hydrogen bond is not observed experimentally. It is neither found by calculation, if energy-optimization is carried out, when the attractive force is minor and the repulsive one major [25].

2.3.1 Triptycenes

Triptycenes (2-15) occupy a peculiar position in this respect. X-ray crystallographic data of the series of triptycenes show that, if there is a substituent at the 9-position of triptycene, the nonbonding distance between the α-atom of the substituent and a *peri*-atom (1, 8, or 13) is ca. 2.7 Å and that between the β-atom of the 9-substituent and the *peri*-atom is ca. 3.0 Å [26]. These distances are smaller than the sum of the van der Waals' radii of two non-hydrogen atoms normally. That means, in every rotational isomer, there is a strong repulsive force and this cannot be avoided by taking another rotational position. This enables a weak attractive interaction to be effective in favoring a particular conformation in which it is possible to exist. There may be another merit in these molecules, that is the overlaps of orbitals concerned are effective because of the proximity.

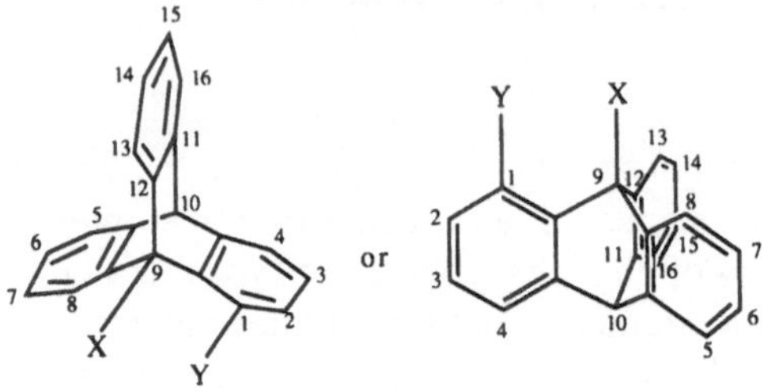

2-15

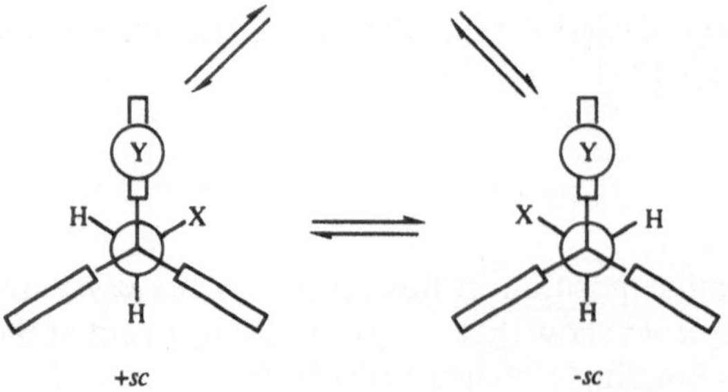

Rotational isomers of the 9-substituted triptycenes can be drawn in the following way using the Newman type projections along the C_9–$C_{substituent}$ bond. Triptycenes are convenient probes for the search for attractive molecular interactions because the population ratios of rotamers are determined easily and interactions between the two groups concerned are very strong due to the proximity of the groups. The stereochemistry of the *ap* and the ± *sc* forms are easily determined. If the 9-substituent is a primary alkyl group, its protons that are attached to the α-carbon of the substituent give an AB quartet pattern in its NMR spectra, if the conformation is *sc*, and make a singlet pattern, if the conformation is *ap*. Furthermore, the freezing of the conformations is easy thanks to high barriers to rotation. It usually suffices to measure the NMR spectra at $-20\,°C$ to see separate signals for the respective isomers.

2.3.1.1 Long Range Coupling in NMR Spectra

Through-space spin-spin coupling in NMR spectra has been reported on various occasions. Triptycenes give decisive information on the mechanism of long range coupling. The methyl protons in 1,2,3,4-tetrafluoro-8,13-dichloro-9-methyltriptycene (2-16) in CD_2Cl_2 give separate signals at $-70\,°C$ at 400 MHz. The ap-hydrogen to the fluoro substituent shows a coupling constant of 8.7 Hz, whereas those at the $\pm sc$ positions have a constant of 6.1 Hz [27]. That is a geometrically remote proton gives a larger coupling constant than those in proximity. Interestingly C_9 of the triptycene skeleton failed to give an observable coupling with the 1-F.

The high degree of coupling between the 1-F and the ap-H must be attributed to the efficient overlap of the orbitals concerned, because C_9 shows no coupling with the 1-fluorine atom. The mechanism of the spin-spin coupling is the transfer of the spin state of a nucleus to another via electrons. Since C_9 of the triptycene does not couple with the 1-fluoro substituent, the information of the 1-F spin must be conveyed through space. The back-lobe of the electron orbital that forms the $ap\text{-H-to-}C_{Me}$ bond can make an efficient overlap with the electron orbitals of the 1-F, because it extends in that direction.

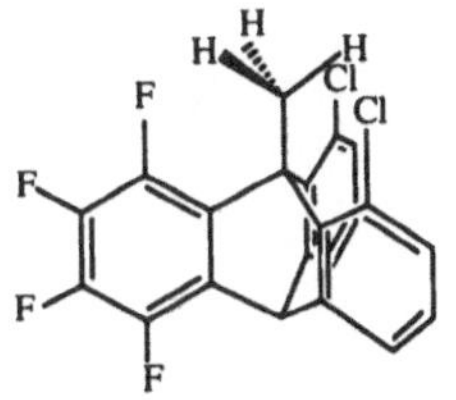

2-16

2.3.1.2 Steric Repulsion

Since the steric congestion in the vicinity of the 9-substituent in triptycenes is very severe, a small difference in the steric size of the substituent has a large effect on the population ratios of rotamers. The most well known example of steric repulsion in controlling the population ratio of rotamers in the triptycene system is the absence of the ap-isomer in 1-substituted 9-isopropyltriptycenes (2-17). In these compounds three rotational isomers are also possible: ap, $+sc$, and $-sc$. However, the presence of the ap-isomer is not observed with the exception of the 1-fluoro compound [28]. The results are easily explained in the following way.

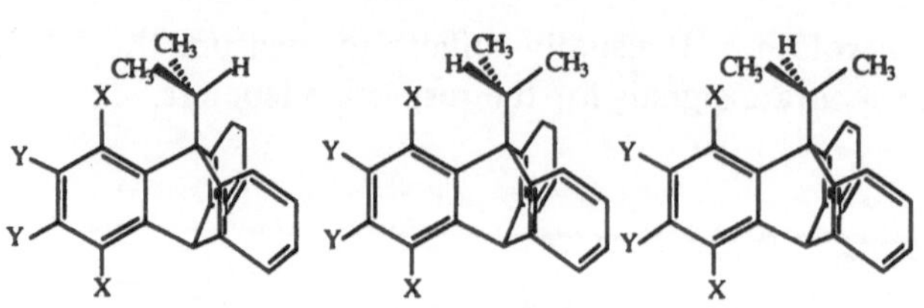

ap-2-17   +sc-2-17   -sc-2-17

If there is a substituent at the 1-position which is bulkier than a hydrogen, then the steric repulsion for the *ap*-isomer would be much larger than that in the *sc*-isomers because the two bulky methyl groups flank the 1-substituent in the conformation. By contrast, in the *sc*-conformations, only one methyl group and a hydrogen atom flank the substituent. In the 1-fluoro compound, due to a small size of the fluoro substituent, the steric repulsion between the fluoro substituent and the two methyl groups seems to be appreciably smaller than in the cases of other substituents, rendering the existence of the *ap*-isomer possible.

These considerations can be applied to other substituents. Indeed, triptycenes (**2-18**) which carry a primary alkyl group at the 9-position and a substituent at the 1-position are known to take the *ap*-conformation preferentially and those (**2-19**) which carry a *tert*-alkyl group also prefer the *ap*-conformation. Due to the steric effect. the *sc*-conformation seldom exceeds the statistical value. Interestingly, the position of the β-non-hydrogen atom in the substituent takes the outer position with respect to the triptycene skeleton, as is shown in **2-19** [29], due to the steric effect again.

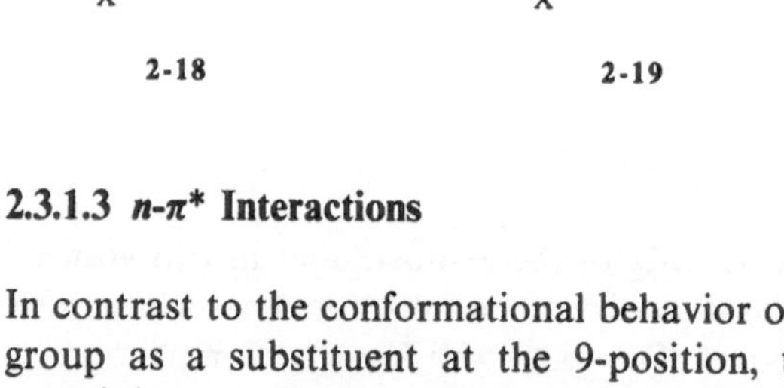

2-18   2-19

2.3.1.3 *n*-π* Interactions

In contrast to the conformational behavior of triptycenes which carry an alkyl group as a substituent at the 9-position, those which carry a substituent containing a carbonyl moiety at the 9-position show an interesting tendency. 1,4-Dimethoxytriptycenes (**2-20**) that carry a 2-substituted 2-oxoethyl group at the 9-position all showed that the *sc*-conformations are stable and no *ap*-forms are detected by NMR spectroscopy, irrespective of the fact that 9-allyl-1,4-dimethoxytriptycene gives a $\pm sc/ap$ ratio of 0.7 [30]. Although the ether–oxygen and the carbonyl interaction is known to be weak [21], it seems ap-2-20 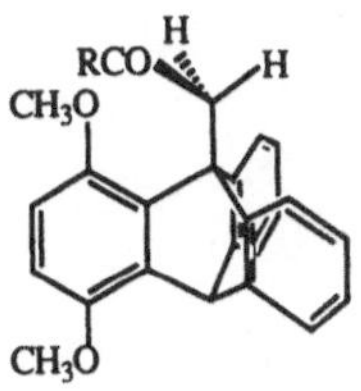   +sc-2-20 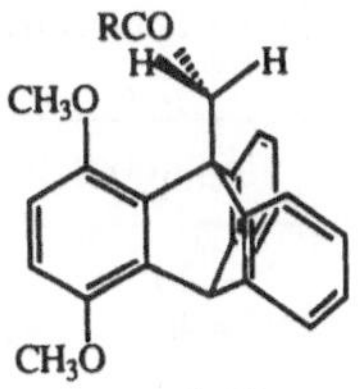   -sc-2-20

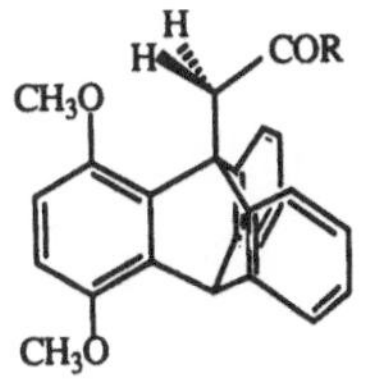

that the stabilization due to that interaction is the key factor which stabilizes the *sc*-conformation.

It is also interesting to note that 9-cyanomethyl-1,4-dimethoxytriptycene **(2-21)** shows the presence of up to 70% *sc*-conformation, whereas the corresponding 9-(2-butynyl) compound **(2-22)** exists almost exclusively as the *ap*-conformation. Since the affinity of the cyano group toward a nucleophile is much smaller than other carbonyl moieties, it is natural that the ether-oxygen and cyano group interaction is weaker than the corresponding oxygen-carbonyl interactions. Yet, it is clear that the cyano group attracts a nucleophile, here it is the ether-oxygen, to make it possible for the *sc*-isomer to be a major component. It is especially so when one compares the results with the acetylenic compound whose steric size should be quite close to that of the cyano compound but whose electrophilic nature is weak.

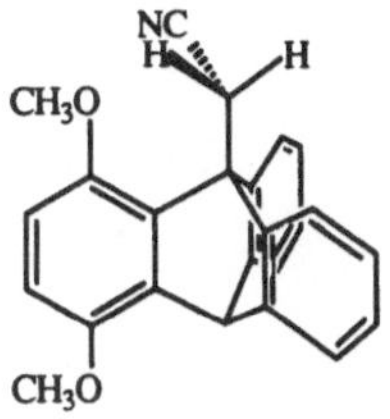

-sc-2-21 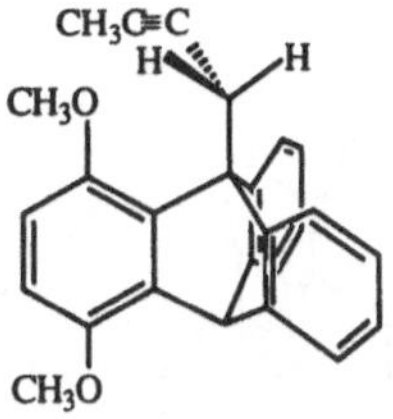   -sc-2-22

The above results indicate that the ether-oxygen is too strong a base to show the sequence of the electrophilic nature of carbonyl moieties by the population ratios of rotamers. Therefore, it is interesting to see the cases of weaker nucleophiles. As an example, a chloro substituent was chosen as the nucleophile. The results obtained with 9-(2-substituted 2-oxoethyl)-1,2,3,4-tetrachlorotriptycenes **(2-23)** are shown in Table 2.2 [30].

As is clear from the table, thanks to the weak basicity of the chloro substituent, one can now see the sequence of the affinity of various carbonyl

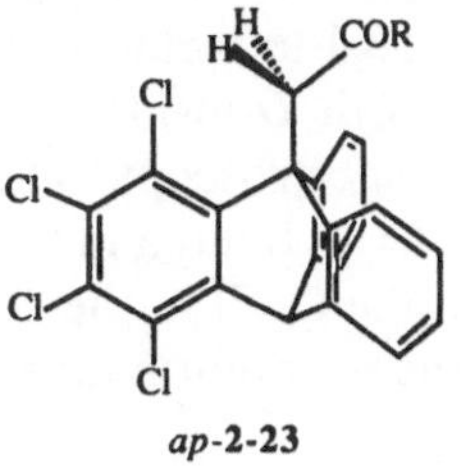

ap-2-23 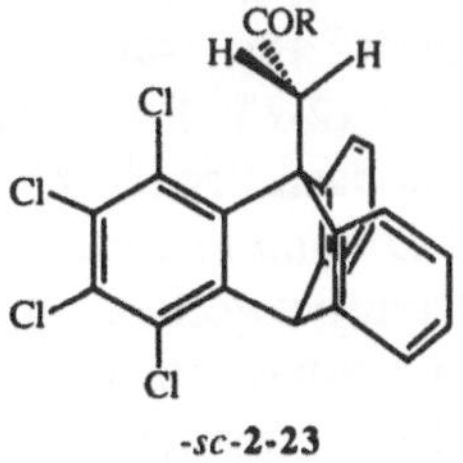   -sc-2-23

Table 2.2. Population ratios in 1,2,3,4-tetrachlorotripty-
cene derivatives (2-23) in $CDCl_3$

Substituent (COR)	$\pm sc/ap$	Temp/°C
$COCH_3$	$\geqslant 10$	-20
$COOCH_3$	6	-40
CHO	1.1	-20
CN	0	-40

groups toward nucleophiles. It is acetyl > methoxycarbonyl > formyl > cyano. This sequence is in accord with what is commonly accepted in organic chemistry, except that the formyl group is less vulnerable to the attack of a nucleophile than normally.

X-ray crystallography of these compounds revealed the cause of this anomalous tendency [31]. That is, the acetyl group takes an O-inside conformation (2-24) at the carbonyl moiety, whereas the formyl group an O-outside (2-25). This is because of the steric effect: of the two groups, oxo and methyl, in the acetyl, the former is smaller than the latter and thus the former is directed toward the triptycene skeleton to minimize the repulsive interactions. By contrast, in the formyl group, the hydrogen is much smaller than the oxo–oxygen and thus the hydrogen is directed toward the triptycene skeleton.

2-24   2-25

Having O-outside conformation, the sc-forms of the formyl compound are disfavored because of the close arrangement of two dipoles in a parallel fashion, whereas the sc-form of the acetyl compound are favored because two dipoles are arranged in an antiparallel fashion. Therefore, the apparent tendency of the electrophilicity of the acetyl and the formyl groups includes this effect and does not disagree with common knowledge of organic chemistry, that is, that the formyl is more electrophilic than the acetyl.

Due to the fact that the oxo-oxygen and the ether-oxygen in the methoxycarbonyl moiety have about the same size, 1,2,3,4-tetrachloro-9-(methoxycarbonylmethyl)triptycene (2-23: R=OCH_3) has both O-inside and O-outside conformations. This has been proved by infrared spectroscopy. The methoxycarbonyl group has a C–O bond in addition to the C=O bond and the former reduces the electric repulsion between the C=O and the C–Cl in the O-outside conformation. This is the reason for the favoring of the sc-conformations in this compound, as compared to the formyl compound.

Another n-π^* type of molecular interaction studied with the use of triptycene rotamers is the charge-transfer interactions in which the ether-oxygen acts as an electron-donor and the benzene ring as an electron-acceptor [32]. From the general tendency of the sc-forms to be relatively unstable in substituted triptycenes, when a 9-substituent is a primary alkyl group, it is judged that 9-benzyltriptycenes (2-26) form exceptions because they give population ratios (sc/ap) of more than 2, which is the statistical value for the conformer equilibration. When the benzyl-phenyl group carries an electronegative group, the sc/ap ratio becomes large and it is small when the substituent in the benzyl group is electron-donating. The results indicate that there is an attractive interaction between the 1-methoxyl group in the sc-form (sc-2-26) and the benzyl group and the benzyl–phenyl group must be the electron acceptor. The last statement is also supported by the fact that, when a less electron-donating substituent is introduced at the 1-position, the sc/ap ratio decreases. It is unusual that the charge-transfer interaction in which benzene or alkylbenzene acts as an electron-acceptor, but it is possible due to the close proximity of the substituents concerned and it is detected by the population ratios of the rotamers because of the favorable situation of the triptycene system.

ap-2-26 z -sc-2-26

2.3.1.4 n-σ^* Interactions

n-π^* Interactions in the nucleophile-carbonyl system are known as an incipient transition state for the carbonyl addition. In the same way, interactions between sulfides and nucleophile were studied by X-ray crystallography and it is claimed that they are a model for the incipient transition state for S_N2 reactions [33]. It is natural to seek this kind of interaction in the triptycene system.

Natural candidates chosen from the experiences with the carbonyl-oxygen or chlorine interactions were those triptycenes which carries a methoxyl group at the 1-position and a chloroalkyl group at the 9-position. However, in these cases, the interaction between the chloroalkyl and the methoxyl groups was too strong to observe population ratios of rotamers, and rather they gave reaction products, namely cyclic ethers [34]. For example, treatment of 9-(2-hydroxyethyl)-1,4-dimethoxytriptycene (2-27) with thionyl chloride gives a cyclic ether (2-28) directly.

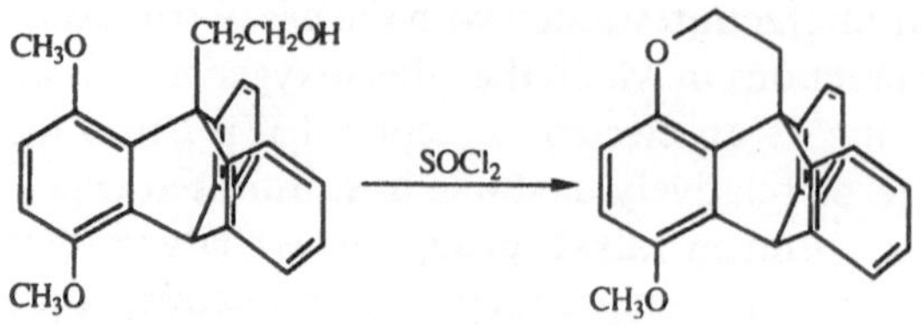

In order to reduce the electron affinity of an electron-accepting group, acyloxymethyl groups are selected. The population ratios of rotamers in 9-(2-acyloxyethyl)-1,4-dimethoxytriptycenes (**2-29**: R=acyl) are listed in Table 2.3, together with a reference compound, 9-(2-methoxyethyl)-1,4-dimethoxytriptycene [35]. The results clearly indicate that the change of the substituent at the β-position of the 9-substituent, from a methoxy to an acyloxy, causes the increase in the number of *sc*-isomers, irrespective of the similar steric situation for all the compounds examined, and the increase is enhanced when the acidity of the carboxylic acid corresponding to the acyl group increases. This is convincing evidence that the acyloxymethyl group accepts electrons from the methoxy-oxygen to stabilize the *sc*-form. Therefore, we think that this type of interactions should be called n-σ* charge-transfer interactions.

This will be the model of incipient transition state of S_N2 reactions which are now known to proceed via an complex of an alkyl halide and a nucleophile from the gas phase study [36] as well as having theoretical support [37].

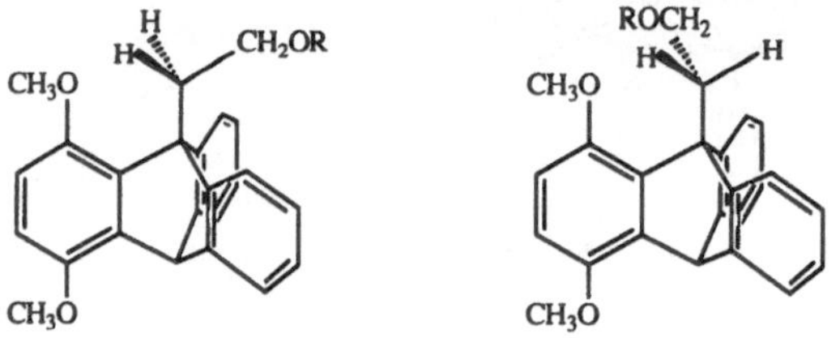

When the electron donor is switched from a methoxy group to a chloro substituent, the population of the *sc*-isomers drastically decreases. 9-(2-

Table 2.3. Population ratios in 9-(2-acyloxyethyl)-1,4-dimethoxytriptycenes (**2-29**) and related compounds in CDCl₃

Substituent (R)	± *sc/ap*	Temp/°C	pKa of ROH
CH₃	0.7	− 50	14.4
CH₃CO	1.5	− 45	4.76
p-CH₃OC₆H₄CO	1.7	− 50	4.47
C₆H₅CO	1.7	− 50	4.20
ClCH₂CO	2.0	− 45	2.86
Cl₂CHCO	2.5	− 45	1.26
Cl₃CCO	3.0	− 50	0.65
F₃CCO	3.0	− 50	0.23

Acetoxyethyl)-1,2,3,4-tetrachlorotriptycene (**2-30**: X=acetoxy) does not show any evidence of the presence of the *sc*-isomer. Even 9-(2-Tosyloxyethyl)-1,2,3,4-tetrachlorotriptycene, which should favor the *sc* conformation by virtue of the stronger electron accepting nature of the tosyloxy group, more than the corresponding acetoxyl compound, showed a practically negligible presence of the *sc*-isomer. However, the interaction between the chloro substituent and a chloromethyl or a tosyloxymethyl group were able to be detected by mass spectroscopy or by a chemical reaction. Electron impact mass spectra of 9-(2-chloroethyl)-1,2,3,4-tetrachlorotriptycene (**2-30**: X=Cl) showed abundant ion peaks at those m/z values corresponding to a loss of a chlorine atom. The chloroethyl compound and the tosyloxyethyl compound were easily hydrolyzed, compared with other ethyl chloride derivatives, affording the corresponding 2-hydroxyethyl compound on chromatography using silica gel. The tosyloxy compound was hydrolyzed even by atmospheric moisture. These facts suggest that a chloronium ion (**2-31**) or the transition state for forming the chloronium ion is stabilized.

2-30   2-31

9-(Chloromethyl)-1,4-dimethoxytriptycene (**2-32**: X=CH$_3$O) shows interesting behavior in its NMR spectra at 60 MHz, when temperature is varied. The signal due to the methylene protons is singlet at room temperature, broadens when temperature is lowered to ca. 0 °C, and sharpens again when the temperature is further lowered. These phenomena are interpreted in the following way [32]. At room temperature, there is a fast equilibrium of two isomers in very different populations. When temperature is lowered, the signal begins to decoalesce. However, cooling causes a decrease in the population of the less stable rotamer below the noise level, thus observation of the isomer becoming impossible at even lower temperatures.

This means that one isomer, which is *ap* according to the NMR spectra at low temperatures, is predominant. This was first attributed to the steric effect [32]. However, the data discussed above strongly indicate that in compound **2-32** (X=CH$_3$O) charge-transfer interactions should be possible between the CH$_2$Cl and the methoxy-oxygen in the *ap* form because the groups concerned are situated close to each other and favorably for the interaction. It might also be the consequence of a favorable head-to-tail arrangement of two dipoles, the methoxyl group and CH$_2$Cl. With this in mind, a detailed study was performed for this type of compound [38].

ap-**2-32**   -sc-**2-32**

At 400 MHz, the compound exhibited the presence of ca. 3% of the $\pm sc$ isomer and 97% of the ap at $-30\,°C$ in CDCl$_3$. When the 4-substituent was changed to hydrogen and a nitro (**2-32**: X=H or NO$_2$) to decrease the electron density on the 1-methoxy group and consequently to increase the ionization potential, the $\pm sc$ population increased. In the less polar solvent, toluene, the population of the $\pm sc$-**2-32** (X=NO$_2$) was ca. 8%. Though this increase in the population ratios of the sc-isomer in the nitro compound could be attributed to the less facile charge-transfer interactions due to the high ionization potential of the ether-oxygen, it could also be attributed to the smaller steric size of the ether-oxygen when its electron density is lowered. The possibility of the head-to-tail arrangement of two dipoles is not ruled out either.

The presence of the interaction between the CH$_2$Cl group and the ether–oxygen was proved by X-ray crystallography of 9-(chloromethyl)-1,4-dimethoxytriptycene and related compounds. This compound (**2-32**: X=CH$_3$O) crystallized in ap conformations and possessed a very unusual structure: the 1-methoxy group was out of the benzene plane to which it is attached, the twisting angle being 35°, whereas the other methoxy group at the 4-position was practically coplanar with the benzene ring. Molecular structures of related compounds reveal that the nonplanarity of the 1-methoxy group is significant. Molecular contact could not be the cause for the distortion of the 1-methoxy group. The structure was interpreted by the assumption that both the ionization potential of the ether-oxygen and the geometry of the molecule become favorable for the charge transfer interactions.

2.3.1.5 Hydrogen Bonding Involving a Methyl Group

Hydrogen bonding is frequent with the OH, NH, and, though weak, SH groups, when an electron-donor is present. The electron-accepting orbital of the X–H bond must be low-lying in order to facilitate the charge transfer interaction.

According to this criterion, the C–H group is a very weak electron-acceptor and thus hardly ever forms a hydrogen bond with electron-donors. Among the C–H bonds, those which are electronegative, such as those in hydrogen cyanide, chloroform and acetylene, are known to form a hydrogen bond [39]. The C–H bond which involves sp^2 carbon is claimed to form hydrogen bond as well [40].

CH$_3$ . . .O Hydrogen Bond

The methyl group, due to its low electronegativity, does not form a hydrogen bond under normal conditions. It has been reported that by calculations even a methyl group in nitromethane and acetonitrile does not form a hydrogen bond with an ether-oxygen [25]. However, there are other reports, which also treat hydrogen bonding involving a methyl group by calculation, and these claim that CH$_3$-π hydrogen bonding is possible [41]. It seems very difficult to demonstrate this kind of interactions due to the fact that the hydrogen bonding involving a methyl group is very weak, if it is at all possible, and the van der Waals' repulsive force is strong. Absence of CH$_3$-O hydrogen bond in the case of nitromethane will be understood in this way.

The 1,9-disubstituted triptycene system is promising for the manifestation of hydrogen bonding involving a methyl group, because the difference in van der Waals' repulsion between the rotational isomers is nearly cancelled out and overlaps of the orbitals concerned should be very efficient due to the close proximity of the two groups concerned. Thus it would be interesting to observe the population ratios of triptycene derivatives which might form a CH$_3$-O or CH$_3$-π hydrogen bond.

2-33   2-34

There have indeed been cases in which a phenomenon could be explained by assuming the presence of a CH$_3$-O hydrogen bond. For example, both the population ratio of the *sc*-conformation over the *ap* in 9-ethyl-1,4-dimethoxytriptycene (**2-33**) [42] and that in 9-allyl-1,4-dimethoxytriptycene (**2-34**) are ca. 0.7 [30], irrespective of the fact that the van der Waals' radius of a methyl group is much larger than half the thickness of a π-system. This could be interpreted as the methyl-oxygen interaction acting attractively to stabilize the *sc* forms in compound **2-33**, whereas such an interaction is not possible in compound **2-34**.

It was a good starting point, therefore, to change the electron-density on the 1-methoxy group of 9-ethyl-1-methoxytriptycene for examination of the population ratios of rotamers. If the electron-density on the 1-methoxy-oxygen is increased, the *sc*-form should be favored due to the interactions and vice versa. This was accomplished by introducing a substituent into the 4-position of the compound.

The results indicated, however, that hydrogen bonding between the methyl group in the 9-ethyl and the 1-methoxy-oxygen was too weak to account for the

ap-**2-35** 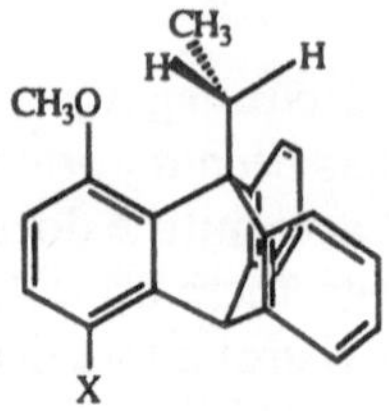   -sc-**2-35**

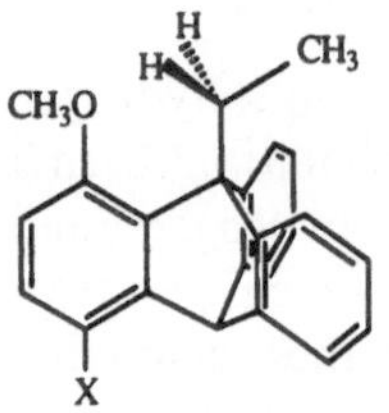

population ratios: decreasing the electron-density on the 1-methoxy-oxygen increased the population of the $\pm$ sc-form rather than the ap. This is explained by the steric repulsion, because the low electron density on the 1-methoxy-oxygen means a small steric size that favors the $\pm$ sc-form. Thus the increase in the population of the sc in the nitro compound (**2-35**: X=NO$_2$) is explained by the steric effect as well as the charge-transfer interactions between the 1-methoxy-oxygen and the methyl group in the 9-ethyl. The results might be due to the weak basicity of an aromatic ether-oxygen and to the weak acidity of an aliphatic methyl group.

The acidity of a methyl group can be increased if it is attached to an aromatic ring and the basicity of an ether–oxygen can be increased by using an aliphatic ether. With these combinations it should be possible to demonstrate the CH$_3$–O hydrogen bond using the population ratios of rotamers. However, the variation in the latter is rather limited and thus aromatic ethers had to be used. Thus 9-(aryloxymethyl)-1,4-dimethyltriptycenes (**2-36**) were designed [43]. These compounds furnish varying basicity of the aryloxy–oxygen and the 1-methyl group must be more acidic than the aliphatic one.

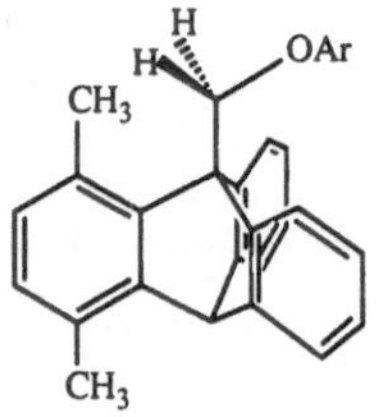

ap-**2-36** 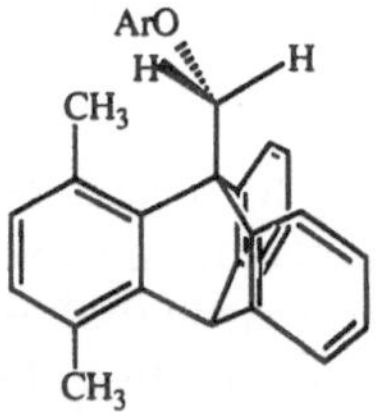   -sc-**2-36**

The results of population study are shown in Table 2.4. Apparently the population of the sc-form increases as the basicity of the ether-oxygen increases. This is a good indication that CH$_3$–O hydrogen bonding exists. However, the Hammett plot here gives a very small negative ρ value. This is due to a delicate balance between repulsive and attractive forces, because increasing the electron density on the ether-oxygen enhances the hydrogen bonding but at the same time increases the repulsive force due to the fact that the effective size of the oxygen increases, thus rendering the CH$_3$–O hydrogen bond less favorable.

This point was nicely demonstrated by the population ratio of rotamers in 9-(phenoxymethyl)-1,4-dimethyltriptycene. It gives the sc/ap value of 0.224,

Table 2.4. Population ratios in 9-aryloxymethyl-1,4-dimethyl-triptycenes (**2-36**) in $CDCl_3$ at $-50\,°C$

Aryl	$\pm sc/ap$[a]	σ constant
4-$(CH_3)_2NC_6H_4$	0.289	− 0.600
4-$CH_3OC_6H_4$	0.242	− 0.268
C_6H_5	0.224	0.000
4-$O_2NC_6H_4$	0.191	0.778
3,4-$(O_2N)_2C_6H_3$	0.165	1.385[b]

[a] The ratios are calculated by dividing the observed population of sc-forms by 2 to compare the stability of a single site of the sc with the ap
[b] This was calculated from the pKa of 3,4-dinitrobenzoic acid [44].

which is very close to that in 9-methoxy-methyl-1,4-dimethyltriptycene, 0.225 [45]. Even with this apparent deviation from the expected which takes into account the basicity of the methoxy-oxygen and the phenoxy-oxygen, the electronic effect in the 9-alkoxymethyl-1,4-dimethyltriptycenes (**2-37**), where the alkoxy is either a methoxy, ethoxy, or 2,2,2-trifluoroethoxy, produces the normal tendency: an electron rich alkoxy-oxygen gives a greater $\pm sc$ population than an electron-poor alkoxy-oxygen.

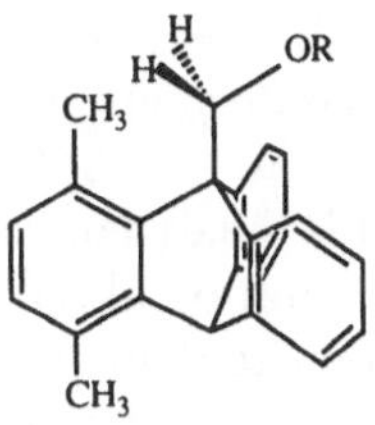

ap-**2-37** 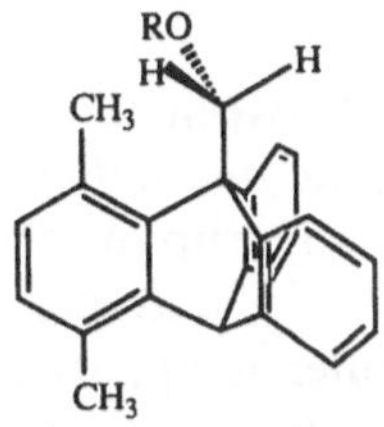   -sc-**2-37**

The apparent low population of the $\pm$ sc-forms in these compounds is the result of the steric repulsion which dominates in the form. If the energy difference due to steric repulsions between the $\pm$ sc-form and the ap can be made negligible, then it should be possible to achieve a situation as in which the $\pm$ sc-form is the major isomer.

This possibility can be examined by taking 9-isobutyl-1,4-dimethoxy-triptycene (**2-38**) as an example [46]. According to the information on the structure of the substituent obtained by X-ray crystallography, the γ-atom of the 9-substituent, which is an X of a XCH_2CH_2 structure, sticks out from the triptycene skeleton. This is reasonable, if one considers the steric effects which are expected when the atom is directed inward relative to the triptycene skeleton. Thus, the molecules will also have such conformations even in solution. However, in this conformation, the two hydrogen atoms attached to the β-atom of the 9-substituent are located very close to the peri-hydrogens of the triptycene system. If either one of the hydrogens is substituted by other atoms, the conformation of the 9-substituent has to change from that which is stable in the molecules mentioned above. 9-Isobutyltriptycene is the case of this.

The β-atom carries two methyl groups here. To avoid steric repulsions between various atoms, the substituent is expected to take a conformation in which one of the methyl groups directs inward and the two methyl groups avoid complete eclipsing with the substituents of the α-position, that is a $\pm ac$-conformation. Such conformations of 9-isobutyl-1,4-dimethoxytriptycene (2-38) are shown in the next structural formula.

ap-2-38   -sc-2-38

(Conformation about the C_α–C_β bond of the 9-substituent is ac.)

These structural features of 9-isobutyltriptycene derivatives will nearly erase the energy difference by repulsion between the $\pm sc$ and ap conformations. Indeed, ^1H NMR spectra of this compound in dichloromethane-d_2 at $-100\,°C$ showed the presence of $\pm sc$-isomer to be up to 66%. Of interest is the dynamic behavior of the isopropyl group in the sc and ap rotamers. The rotation about the C_{i-pr}–CH_2 bond in the sc-isomer is apparently slower than that in the ap-isomer: the NMR signal due to the methyl protons in the sc-isomer is a distinct doublet which further couples with the methine proton, while that in the ap is a broad signal to indicate that the rotation takes place on the NMR time scale in the ap isomer. This will mean that there is a deeper energy well in the sc-isomer than in the ap and suggests that the CH_3–O interaction is responsible for this phenomenon.

In addition to these two lines of evidence that support the presence of the CH_3–O hydrogen bond, X-ray crystallographic analysis also showed an interesting structural feature that supports the presence of such an interaction. Namely, in the crystal, one of the methyl groups in the substituent, that is directed inward relative to the triptycene skeleton, is very close to 1-methoxy-oxygen. The 1-methoxy group is out of the plane of the benzene ring, to which it is attached, to an extent of $10°$, whereas the 4-methoxy group is practically coplanar with the benzene ring. The structure of the 1-methoxy group must be the consequence of a demand to lower the ionization potential of the oxygen atom.

$CH_3 \ldots \pi$ Hydrogen Bond

After establishing the presence of the CH_3–O hydrogen bond, one is also interested in doing the same with the CH_3-π hydrogen bond. Being a weaker base than an ether-oxygen, it was doubted whether the π-base could form a hydrogen bond, though acetylenes definitely form a CH-π bond [47]. There were some papers which reported the presence of such interactions [48], but they were doubtful at best.

The triptycene system should serve to look for such interactions shown by the population ratio of rotamers. 8,13-Dichloro-9-(substituted benzyl)-1,4-dimethyltriptycenes (2-39) were selected as candidates to show such possibilities [49]. Here, the 1-methyl group acts as an electron acceptor and the benzyl-benzene ring as an electron-donor. The chlorine substituents were thought convenient due to their steric sizes, because the steric sizes of the methyl group and a chloro substituent are not very different from each other and the energy differences due to steric repulsions between the rotational isomers could be kept to a minimum. Increasing the electron density of the benzyl-benzene ring will increase the population of the sc-isomers if such interactions are present. Indeed, such an electronic effect on the population ratios was observed when substituents, NO_2, H, and $N(CH_3)_2$, were introduced at the p-position of the benzyl group (2-39).

ap-2-39   +sc-2-39

However, there could be an argument against accepting these experimental results as evidence for the presence of the CH_3-π hydrogen bond. The argument is that charge-transfer interactions between the chloro substituent and the benzyl-benzene ring are possible in the ap-conformation and such an interaction is enhanced if the benzyl-benzene ring carries an electronegative group. Although it is unlikely that the charge transfer interactions are the main factor that stabilizes the ap-form, because of the high ionization potential of the chloro substituent and because of the fact that, even in the sc-conformation, such an interaction is possible to a certain extent, it is not possible to completely rule out that this is so.

To avoid this argument, it would be wise to keep the electron-donating ability of the π-system constant and to change the acidity of the 1-methyl group. This is accomplished by introducing a substituent at the 4-position of triptycene (2-40). The results are shown in Table 2.5, together with some relevant data.

Table 2.5. Population ratios of rotational isomers in 4-substituted 9-benzyl-8,13-dichloro-1-methyltriptycenes (**2-40**) in $CDCl_3$ and calculated difference in total energies by the MM2 method

Subst (X)	± sc/ap	σ_p Const	ΔE (ap-sc)[a]
CH_3	2.30 ± 0.10	− 0.0370	1.077
H	2.22 ± 0.06	0.00	1.008
$COOCH_3$	4.09 ± 0.16	0.385	0.984[b]
CN	3.42 ± 0.14	0.674	1.052

[a] in kcal/mol
[b] Calculation was carried out using the acetyl group due to unavailability of the parameters of the methoxycarbonyl group.

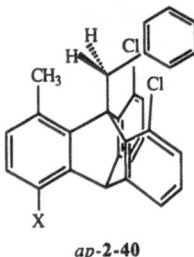

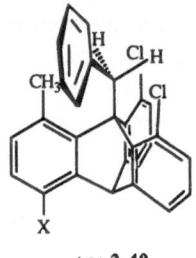

ap-**2-40** $+sc$-**2-40**

As is clear from the data in Table 2.5, increasing the acidity of the 1-methyl group by introducing an electronegative substituent at the 4-position increases the population of the sc-isomer. These results are best understood if one assumes the presence of the CH_3-π interactions where the methyl group is the electron-acceptor and the π-system an electron-donor. The results of molecular mechanics calculations are of interest. Although this method is known to reproduce stable geometries of triptycene derivatives, the prediction made by the calculation is that the sc-forms are always less stable than the ap-forms by ca. 1 kcal/mol. This failure to predict stable conformers is due to the fact that the molecular mechanics calculation does not take into account the charge-transfer interactions. As a corollary, the experimental results indicate that there are charge-transfer type attractions between the methyl group and the π-system in this case.

There are some reports that reveal a conformation in which a methyl group and a π-system are close to each other is favored [50], irrespective of the fact that van der Waals' repulsion disfavors such a conformation. This can be understood by taking the CH-π interactions into account.

2.3.2 9-Arylfluorenes

Though less effective than the triptycene system, the 9-arylfluorene system is also useful for detecting certain molecular interactions. This is due to the fact that, in two conformations possible for 9-arylfluorenes (**2-41**) the substituent

(Y in *sp* or X in *ap*) which is placed above the fluorene ring undergoes strong molecular interactions with the π-system, whereas such interaction is weak in the other position (X in *sp* and Y in *ap*) which is close to the 9-H of the fluorene. If this interaction is repulsive, it means that a bulky substituent tends to take a position that is close to the 9-H. It is clearly shown by the populations of rotamers in the 9-(2-substituted phenyl)fluorene series [51].

If, therefore, one finds an anomalous population ratio from the stand point of the steric effect, then there is a good chance of finding a molecular interaction. One can expect to observe interactions involving the π-system of the fluorene and the substituent on the aryl ring in this series of compounds. The drawback of this system lies in the difficulty of changing the substituent. Therefore, interactions involving the π-system are the only ones that can be found in this system.

sp

ap

if X precedes Y in the sequence rule

2 - 41

2.3.2.1 XH-π Interactions

Hydrogen bonding in which the π-system acts as an electron-donor is known [52]. Therefore, if X or Y in compound **2-41** carries a hydroxyl group or others that can accept electrons, the hydrogen bond exists between the fluorene ring and the group which is above the ring. This kind of interaction was first recognized by observing the rotamer equilibrium in 9-(2-hydroxy-4,6-dimethyl-phenyl)fluorene (**2-42**) [53].

The compound showed a population ratio, *ap/sp*, of 1.80. This value is anomalous in that the corresponding methyl ether gives the same population ratio of 1/3.30. Namely, in the methyl ether, the *sp* form is more stable than the *ap*, whereas the tendency is reversed in the phenol, irrespective of the fact that both compounds carry an oxygen atom and the difference in the structure is located far from the site of steric repulsion.

The stability of the *ap*-phenol, relative to the *sp*, is attributed to the presence of OH-π interactions that are not possible in the *sp*, because the infrared

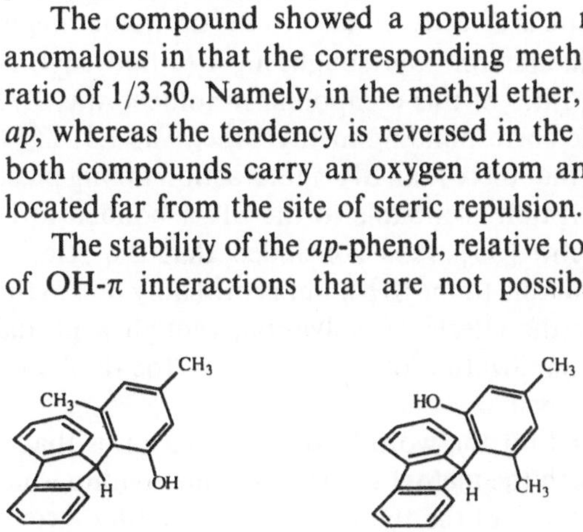

sp-**2-42**

ap-**2-42**

spectrum of the *ap* form showed an O–H stretching band at 3546 cm^{-1} in carbon tetrachloride, corresponding to the OH-π bonded form, whereas the *sp* showed a single absorption band ascribable to the free form at 3611 cm^{-1} in the same solvent. No free form for the *ap* was detected by infrared spectra.

Similarly, 9-(2-hydroxy-1-naphthyl)fluorene gives an *ap/sp* ratio of 2.30, whereas the methyl ether gives one of 1/3.56. Infrared spectra also indicated that the *ap* exists as OH-π bonded form exclusively, whereas the *sp* form is always free.

The case of 9-(2-hydroxymethyl-1-naphthyl)fluorene (**2-43**) is a little different from that of the phenols mentioned above [54]. This compound gives an *sp/ap* ratio of 1.80 in hexachlorobutadiene at 114 °C. Namely, in this compound the *sp* form is more stable than the *ap*. However, this value should be compared with those of 9-(2-methoxymethyl-1-naphthyl)fluorene, which gives an *sp/ap* of 2.25 at the same temperature. The *ap* form of the alcohol is relatively more stable than the corresponding methoxy compound. This situation is again reflected in the infrared spectra. Whereas the *sp* form gives only an absorption due to the free OH, the *ap* form gives two bands in carbon tetrachloride: free at 3630 and bonded at 3580 cm^{-1} with similar intensities. The results are attributed to both the weak acidity of the alcohol and the restriction of freedom of rotation with the OH-π bond formation.

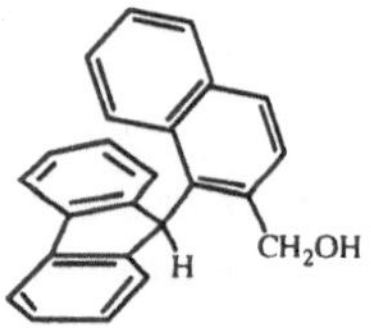

sp-**2-43** 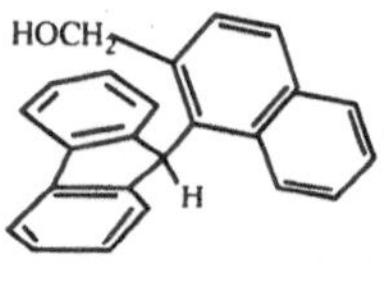   *ap*-**2-43**

The results shown above indicate that there is a better chance of observing weak interactions if a π-accepting site is directly bonded to the aromatic ring in 9-arylfluorenes. Thus an SH-π interaction was sought in 1-(9-fluorenyl)-2-naphthalenethiol (**2-44**). The *sp/ap* ratio of this compound in toluene was 3.72 at 101 °C [55], whereas that for the corresponding methyl ether was 7.31 [56]. This may be taken as an indication that there is an SH-π interaction taking place in the *ap*-form. However, no absorption ascribable to the SH-π bonded form was detected by the infrared spectra [55]. It is concluded that the relative stability of the *ap* form is not attributable to the SH-π but is probably due to the "size" of the methylthio group or the effects on solvation, though a partial contribution of the SH-π bond in stabilization of the *ap* form of the thiol with respect to the methyl thioether is possible.

Quotation marks have been used for the word "size". The reason is that if one considers conformations in which the methyl group of the methylthio is far from the fluorene ring, the effective size of the thiol and the methylthio group must be the same. However, by limiting conformations in some way, the

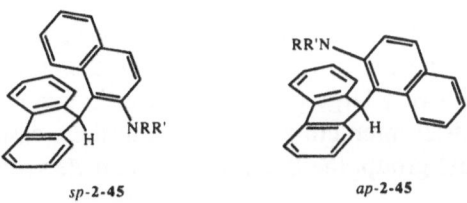

sp-2-44   ap-2-44

methylthio group is contributing to the decrease in entropy. This should be unfavorable for that conformation which, here, is the *ap* form.

The NH group is known to form a stronger hydrogen bond than the SH group. However, the presence of the NH-π bond was inconclusive at best, because infrared spectra did not show the low-frequency shift typical for the bonded form [57]. There should be a better chance to observe such an interaction in the 9-arylfluorene series.

9-(2-Amino-1-naphthyl)fluorene (**2-45**: R=R′=H) and its *N*-methyl derivatives were prepared and their rotamer populations examined [55]. The results show the following *sp/ap*, ratios: R=R′=H 0.52, R=CH$_3$, R′=H 0.44, R=R′=CH$_3$ 6.31. Here again, the preference of rotational isomers is reversed when NH is absent from what it is when an NH group is present. The cause of this phenomenon may be attributed to the presence of the NH-π interactions in the *ap* form which stabilize the conformation.

Interestingly, N–H stretching spectra of these compounds show no significant shift, both for the primary amine and the secondary amine. Namely both *sp* and *ap* forms absorb at about the same frequency (3496 and 3405 cm^{-1} for the primary amine and 3460 cm^{-1} for the secondary amine). Because of its weak nature, the NH-π hydrogen bond does not seem to affect the N–H stretching frequency.

sp-2-45   ap-2-45

The preference of the *ap* form over the *sp* in mobile 9-(2-aminophenyl)-fluorene derivatives (**2-46**) is also recognized. The parent compound (**2-46**: R=H) was found to exist as a mixture of 4:1 *ap* and *sp* isomers at − 51 °C [58], whereas the monomethyl derivative (**2-46**: R=CH$_3$) as a 2.1:1 mixture of the isomers at − 30 °C [59]. This is again abnormal, if one considers that the bulky substituent in the aryl group in 9-arylfluorenes tends to take the *sp* position to avoid steric repulsion. Although they showed the normal N–H stretching frequencies even for the *ap* form, as was the case for compound **2-45**, the ^{1}H NMR spectra of the rotamers (**2-46**: R=CH$_3$) was of interest. The methyl-proton signal of the *ap*-form was a doublet due to the coupling with the

NH proton, whereas it was a singlet for the *sp*. Apparently, the NH exchange is fast in the *sp*, whereas it is slow in the *ap*. This is also evidence for the presence of NH-π bonding: due to the formation of the bond, the exchange of the NH proton becomes slow to make the coupling with the *N*-methyl protons apparent.

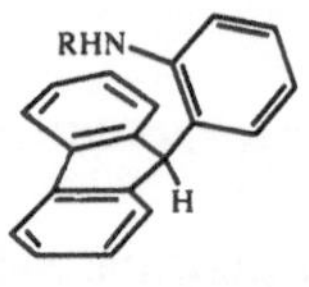

ap-2-46 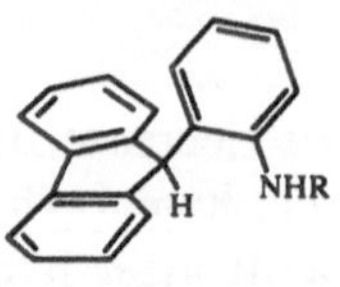   *sp*-2-46

2.3.2.2 π–π^* Interactions

Benzoate of 9-(2-hydroxy-1-naphthyl)fluorene (**2-47**: Ar=C_6H_5) shows an interesting tendency in the population ratio of rotamers [60]. The population ratio, *ap/sp*, is 0.82. This value is much higher than that for the corresponding methoxy compound mentioned above, 1/3.56, though the steric requirement is very similar for both the compounds or even less favorable for the *ap* form of the benzoate than the *ap* of the methyl ether on steric grounds.

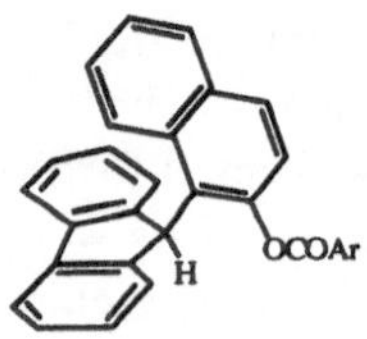

sp-2-47 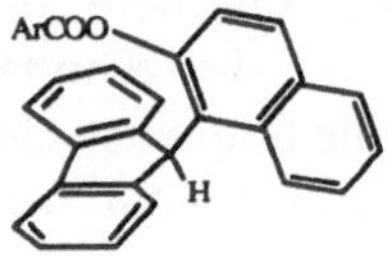   *ap*-2-47

In order to see the cause of this anomaly, a substituent was introduced to the *p*-position of the benzoyl group. The results are shown in Table 2.6. Interestingly, the nitro compound showed a very large population ratio in favor of the *ap* form. It means that charge-transfer interactions take place in these cases between the fluorene ring and the aroyl group, the latter acting as an electron-acceptor and the former as a donor.

Table 2.6. Population ratios in *p*-substituted benzoates (**2-47**) of 9-(2-hydroxy-1-naphthyl)fluorene in CDCl$_3$ at 59 °C

Substituent	ap/sp
CH$_3$O	0.83 ± 0.03
CH$_3$	0.82 ± 0.03
H	0.90 ± 0.03
NO$_2$	2.9 ± 0.2

The structure of the *p*-nitrobenzoate may be written in the following way (**2-48**). This type of conformation in esters is known unfavorable due to the dipole moments arranged in parallel fashion [61]. However, the charge-transfer interactions seem to be strong enough to manifest the preference of the *ap* conformation.

2-48

It is, of course, expected that the charge-transfer interaction will be affected by the steric effect very strongly, because of its particular structure as well as the congested state of the substituent of the naphthalene ring in the *ap* conformation. This expectation is substantiated by the *ap/sp* population ratio found, namely 0.67 ± 0.03 for *o*-methylbenzoate (**2-47**: Ar=*o*-$CH_3C_6H_4$).

2.4 References

1. Le Févre CG, Le Févre RJW, Roper R, Pierens RK (1960) Proc Chem Soc 117
2. Green FD, Chu C-C, Walia J (1964) J Org Chem 29: 1285
3. Hall JR, Wilson MK (1966) Spectrochim Acta 22: 1729
4. Lister DG, Macdonald JN, Owen NL (1978) Internal rotation and inversion, Academic, London
5. Imashiro F, Terao T, Saika A (1979) J Am Chem Soc 101: 3762
6. Siddall TH III, Stewart WE (1969) J Org Chem 34: 233
7. Nair PM, Roberts JD (1957) J Am Chem Soc 79: 4565
8. Yamamoto G, Ōki M (1975) Bull Chem Soc Jpn 48: 3686
9. Yamamoto G, Suzuki M, Ōki M (1983) Bull Chem Soc Jpn 56: 809
10. Ōki M, Matsumoto Y, Toyota S (unpublished work)
11. Pitzer KS (1940) Chem Rev 27: 39. Ito K (1953) J Am Chem Soc 75: 2430
12. Carter RE, Nilsson B, Olsson K (1975) J Am Chem Soc 97: 6155
13. Goldfinger P, Martens G (1961) Trans Faraday Soc 57: 2220
14. Mizushima S (1954) Structure of molecules and internal rotation, Academic, New York, p 41: Mizushima S, Morino Y, Watanabe I, Simanouti T, Yamaguchi S (1949) J Chem Phys 17: 591. Bernstein HJ (1949) J Chem Phys 17: 258
15. Allinger J, Allinger NL (1958) Tetrahedron 2: 64
16. Flett MSTC (1957) Spectrochim Acta 10: 21
17. Lemieux RU, Koto S (1974) Tetrahedron 30: 1933
18. a) Delongchamps P (1983) Stereoelectronic effects in organic chemistry, Pergamon, Oxford. b) Kirby AJ (1983) The anomeric effect and related stereoelectronic effects at oxygen, Springer, Berlin Heidelberg New York
19. ref. (16b), p 48
20. Leonard NJ (1956) Rec Chem Prog 17: 243
21. Leonard NJ, Brown TL, Milligan TW (1959) J Am Chem Soc 81: 504
22. Leonard NJ, Milligan TW, Brown TL (1960) J Am Chem Soc 82: 4075
23. Bürgi HB, Dunitz JD, Shefter E (1974) Acta Crystallogr Sect B 30: 1517: Kaftory M, Dunitz JD (1975) Acta Crystallogr Sect B 31: 2917
24. Kitaura K, Morokuma K (1976) Int J Quant Chem 10: 325

25. Kumpf RA, Damewood JR Jr (1988) J Chem Soc, Chem Commun 621
26. Mikami M, Toriumi T, Konno K, Saito Y (1975) Acta Crystallogr Sect B 31: 2474
27. Yamamoto G, Ōki M (1985) Tetrahedron Lett 26: 457. (1990) Bull Chem Soc Jpn 63: 3550
28. Suzuki F, Ōki M, Nakanishi H (1974) Bull Chem soc Jpn 47: 3114. Yamamoto G, Ōki M (1983) Bull Chem Soc Jpn 56: 2082
29. X-ray crystallography of this type of compounds shows that these are the cases: Toyota S, Ōki M (unpublished work) See also Ref 32 for conformations in solution
30. Ōki M, Izumi G, Yamamoto G, Nakamura N (1982) Bull Chem Soc Jpn 55: 159
31. Nogami N, Ōki M, Sato S, Saito Y (1982) Bull Chem Soc Jpn 55: 3580
32. Suzuki F, Ōki M (1975) Bull Chem Soc Jpn 48: 596
33. Rosenfield RE, Parthasarathy R, Dunitz JD (1977) J Am Chem Soc 99: 4860. Britton D, Dunitz JD (1980) Helv Chim Acta 63: 1068
34. Izumi G, Hatakeyama S, Nakamura N, Ōki M (1981) Bull Chem Soc Jpn 54: 258
35. Izumi G, Yamamoto G, Ōki M (1981) Bull Chem Soc Jpn 54: 3064
36. Olmstead WM, Brauman JI (1977) J Am Chem Soc 99: 4219. Pellerite MJ, Brauman JI (1980) J Am Chem Soc 102: 5993
37. Kozaki T, Morihashi K, Kikuchi O (1989) J Am Chem Soc 111: 1547
38. Tamura Y, Takizawa H, Yamamoto G, Ōki M, Murata S (1990) Bull Chem Soc Jpn 63: 2555
39. Green RD (1974) Hydrogen bonding by C-H groups, Wiley New York
40. Taylor R, Kennard O (1982) J Am Chem Soc 104: 5063
41. Abe K, Hirota M, Morokuma K (1985) Bull Chem Soc Jpn 58: 2713. Takagi T, Tanaka A, Matsuo S, Maezaki H, Tani M, Fujiwara H, Sasaki Y (1987) J Chem Soc Perkin Trans II 1015
42. Nakanishi H, Yamamoto O (1978) Bull Chem Soc Jpn 51: 1777
43. Tamura Y, Yamamoto G, Ōki M (1987) Bull Chem Soc Jpn 60: 1781
44. Dippy JF, Hawkins BP, Smith BV (1964) J Chem Soc 154
45. Tamura Y, Yamamoto G, Ōki M (1987) Bull Chem Soc Jpn 60: 3789
46. Ōki M, Takiguchi N, Toyota S, Yamamoto G, Murata S (1988) Bull Chem Soc Jpn 61: 4295
47. Brand CD, Eglinton G, Tyrrell J (1965) J Chem Soc 5914
48. Nishio M, Hirota M (1989) Tetrahedron 45: 7201. Ōki M (1990) Acc Chem Res 23: 351
49. Nakai Y, Inoue K, Yamamoto G, Ōki M (1989) Bull Chem Soc Jpn 62: 2923
50. Kikuchi H, Hatakeyama S, Yamamoto G, Ōki M (1982) Bull Chem Soc Jpn 54: 3832. Yamamoto G, Ōki M (1984) Bull Chem Soc Jpn 57: 2219
51. Nakamura M, Nakamura N, Ōki M (1977) Bull Chem Soc Jpn 50: 2986
52. Ōki M, Yoshida T (1971) Bull Chem Soc Jpn 44: 1336 and earlier papers of the series
53. Nakamura M, Ōki M (1980) Bull Chem Soc Jpn 53: 3248
54. Saito R, Ōki M (1982) Bull Chem Soc Jpn 55: 2508
55. Moriyama K, Nakamura N, Nakamura M, Ōki M (1987) Gazz Chim Ital 117: 655
56. Moriyama K, Nakamura M, Nakamura N, Ōki M (1989) Bull Chem Soc Jpn 62: 485
57. Ōki M (1960) Kagaku no Ryouiki Zokan No. 40, 1 (1961) Chem Abs 55: 17210e
58. Tukada H, Iwamura M, Sugawara T, Iwamura H (1982) Org Magn Reson 19: 78
59. Nishida A, Takeshita M, Fujisaki S, Kajigaeshi S (1988) Bull Chem Soc Jpn 61: 3919
60. Nakamura M, Ōki M (1976) Chem Lett 651
61. Marsden RJB, Sutton LE (1936) J Chem Soc 1383. LeFévre RJW, Sundaram A (1962) J Chem Soc 3904. O'Gorman JM, Schand W Jr, Schmacker V (1950) J Am Chem Soc 72: 4222. Wilmshurst JK (1957) J Mol Spectrosc 1: 201

3 Barriers to Rotation

3.1 Estimation of Barriers to Rotation

Barriers to rotation can in principle be determined by various methods. However, since we are interested in rather high barriers, the methods that can be used are limited. Those which are useful in this context are the dynamic NMR method and the classical method that uses the analysis of each rotational isomers by various methods. Since the classical method is considered to be measuring the reaction rates that are observed in the isomerization of rotational isomers, it is nothing but a kinetic measurement.

Dynamic NMR is a different method in principle, but is now known to be reliable. The most popular method is the analysis of the line shapes but other methods such as saturation transfer and 2D-NMR measurement may be used. As a simple method of line shape analysis, the coalescence of two or more signals is often observed and free energy of activation for rotation can be obtained from the observed data. Here the methods will be outlined and their reliability will be discussed.

3.1.1 Coalescence Method

When two sites of the magnetic nuclei are exchanging, one should take into account the following. If the exchange is very slow on the NMR time scale, the nuclei at the different sites give rise to two separate signals. However, if the exchange rates are too high on the NMR time scale, those nuclei give rise to an averaged signal that appears at one site. The "NMR time scale" refers to the range of the lifetimes or the rates of exchange, compared with the difference in frequencies of two signals. If the rates of exchange are too high on the NMR time scale, NMR spectroscopy cannot identify the two signals and gives an averaged signal. The critical rate ranges from 10^0 to 10^3 s^{-1} depending upon the chemical shift differences of the two signals.

An example is N,N-dimethylformamide (**3-1**) [1]. Here, the rotation about the C_{co}–N bond is slow at room temperature on the NMR time scale and two methyl protons give signals at different chemical shifts at A and B. At high temperatures, the rotation becomes too fast to see separate signals for the two methyl protons by the NMR technique. Thus a singlet signal ascribable to the averaged of the two (A and B) is now observed.

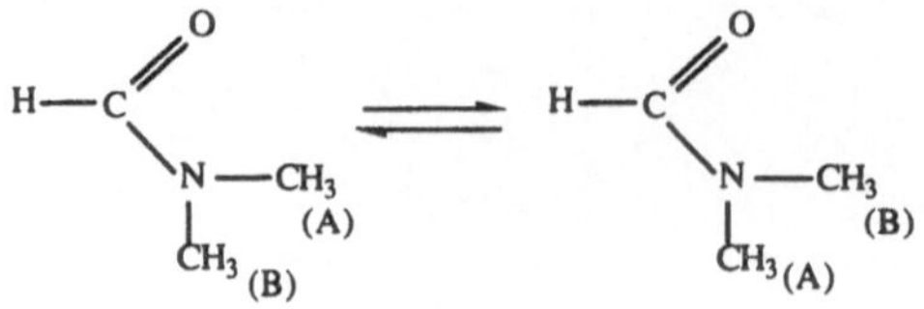

3-1

At the intervening temperatures, the methyl protons give somewhat broadened signals, as shown in Fig. 3.1, and as the temperature is raised the broadening increases and finally a point is reached where no minimum is observed between the two signals. This is called coalescence and the temperature where coalescence of two signals is observed is the coalescence temperature. Above the coalescence point, the signal begins to sharpen until it finally becomes a sharp singlet. The reverse of this phenomenon may be observed. That is, if temperature is lowered, a singlet which is an averaged signal of two may become broad, then passes a coalescence point, and finally splits into two. This is called decoalescence by some scientists. The range of temperatures where the exchange of the magnetic sites of nuclei is slow enough on the NMR time scale to allow observation of two signals is called a slow exchange limit, whereas that where the exchange is fast to give an averaged single signal is called a fast exchange limit.

The coalescence temperature (T_c) expressed by Kelvin and the chemical shift difference between the two signals (Δv) give the rate constant (k_c) of rotation or free energy of activation for rotation at the temperature by the following equations [2].

$$k_c = \frac{\pi}{\sqrt{2}} \Delta v \tag{1}$$

$$\Delta G_c^{\neq} = 4.57\, T_c [9.97 + \log_{10} \frac{T_c}{\Delta v}] \tag{2}$$

These relationships are obtained by solving the Bloch equation, because at the coalescence temperature the mean life time (τ) and the chemical shift difference are in the following relationship (Eq. 3) at the coalescence temperature and the mean life time and rate constants are related as in Eq. (4).

$$\tau = \frac{\sqrt{2}}{2\pi\Delta v} \tag{3}$$

$$\tau = \frac{1}{2} k \tag{4}$$

Strictly speaking, this equation applies to the cases where populations of both A and B are equal, as it is in N,N-dimethylformamide. However, it is also applied to unequally populated cases as well, with a good approximation. An example is N-benzyl-N-methylformamide (3-2). Here, the populations of E and

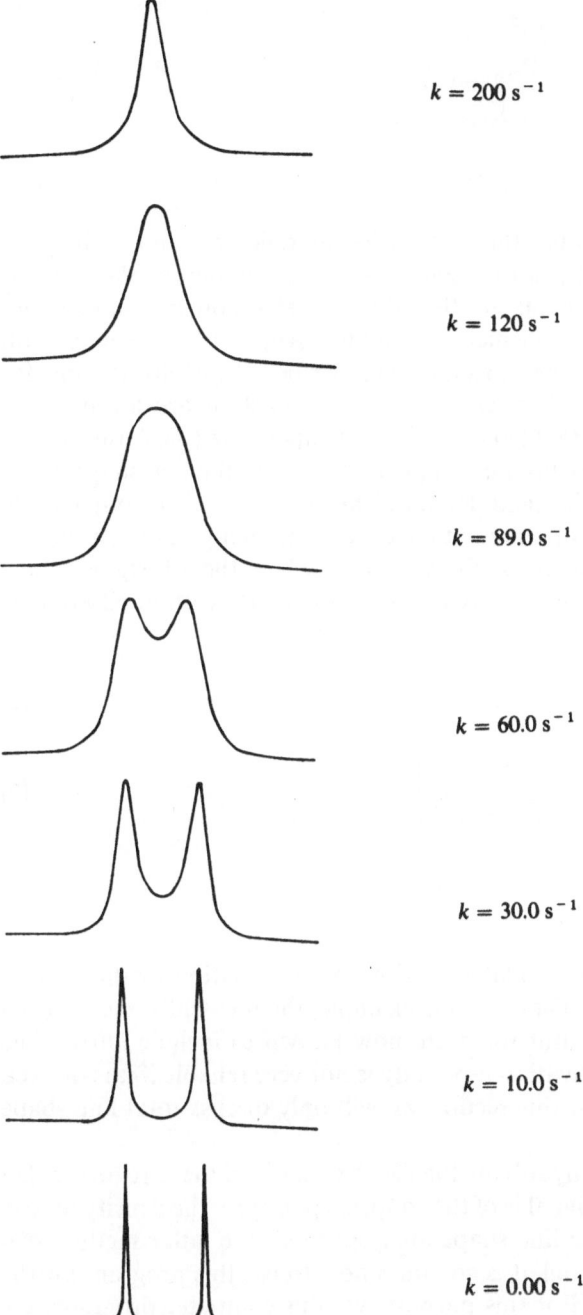

$k = 200 \text{ s}^{-1}$

$k = 120 \text{ s}^{-1}$

$k = 89.0 \text{ s}^{-1}$

$k = 60.0 \text{ s}^{-1}$

$k = 30.0 \text{ s}^{-1}$

$k = 10.0 \text{ s}^{-1}$

$k = 0.00 \text{ s}^{-1}$

Fig. 3.1. Lineshapes of uncoupled AB at various exchange rates: The chemical shift difference is 40 Hz and T_2 0.20 s. (The area surrounded by the curve is magnified to illustrate the feature)

(Z)-3-2   (E)-3-2

Z isomers are not equal because their stabilities are different from each other. The intensities of the methyl-proton signals (A and B), as well as those of the benzylic methylene-proton signals are thus different at a slow exchange limit.

When the signal is of the type which is a coupling AB, both the chemical shift difference (Δv) and the coupling constant (J) must be taken into account for considering the free energy of activation. The relationships are given in the following equations (Eqs. 5 and 6) at coalescence temperature [3]. Although the line shapes are more complex than uncoupled AB signals, the line shape at the coalescence temperature is the same for both the coupled and uncoupled AB: namely, there is no structure observed at the coalescence temperature, as is seen in Fig. 3.2. More complex signals such as the case where the AB signals show further coupling with other nuclei may be treated similarly with good approximations.

$$k_c = \frac{\pi}{\sqrt{2}}\sqrt{\Delta v^2 + 6J^2} \tag{5}$$

$$\Delta G_c^{\neq} = 4.57 T_c \left[9.97 + \log_{10}\frac{T_c}{\sqrt{\Delta v^2 + 6J^2}} \right] \tag{6}$$

3.1.2 Line Shape Analysis

Before good computers were available, various approximation methods were used for line shape analysis. They are, for example, the line width method [4] and hill-to-valley ratios [5]. But these are now known to include fairly large errors and the entropy of activation especially is not very reliable if it is derived by this method. Therefore, in this section we will only discuss total line shape analysis.

Line shapes are analyzed by solving the Bloch equation if they are uncoupled line signals. However, if the signal is of the coupled spins type, the density matrix method must be used for the line shape analysis. Since the latter method can treat the uncoupled case as well, it is common now to use this program for the calculation of spectral lines. For this purpose, various computer programs are now available. Typical examples are DNMR3 [6] and DNMR5 [7]. The former can be used on personal computers but the latter requires a large computer.

For the calculation of line shapes, necessary parameters are chemical shift differences of the signals concerned, coupling constants, and spin-spin relaxation

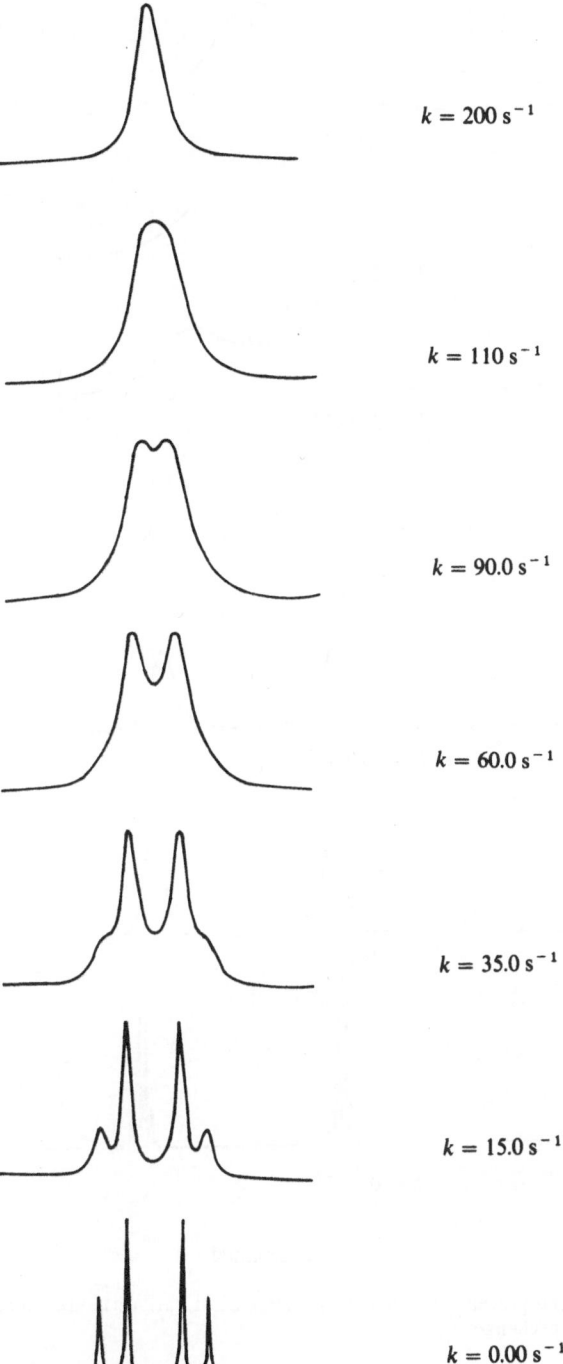

$k = 200 \text{ s}^{-1}$

$k = 110 \text{ s}^{-1}$

$k = 90.0 \text{ s}^{-1}$

$k = 60.0 \text{ s}^{-1}$

$k = 35.0 \text{ s}^{-1}$

$k = 15.0 \text{ s}^{-1}$

$k = 0.00 \text{ s}^{-1}$

Fig. 3.2. Lineshapes of coupled AB at various exchange rates: The chemical shift difference is 40 Hz, coupling constants 14 Hz, and T_2 0.20 s. (The area surrounded by the curve is magnified to illustrate the feature)

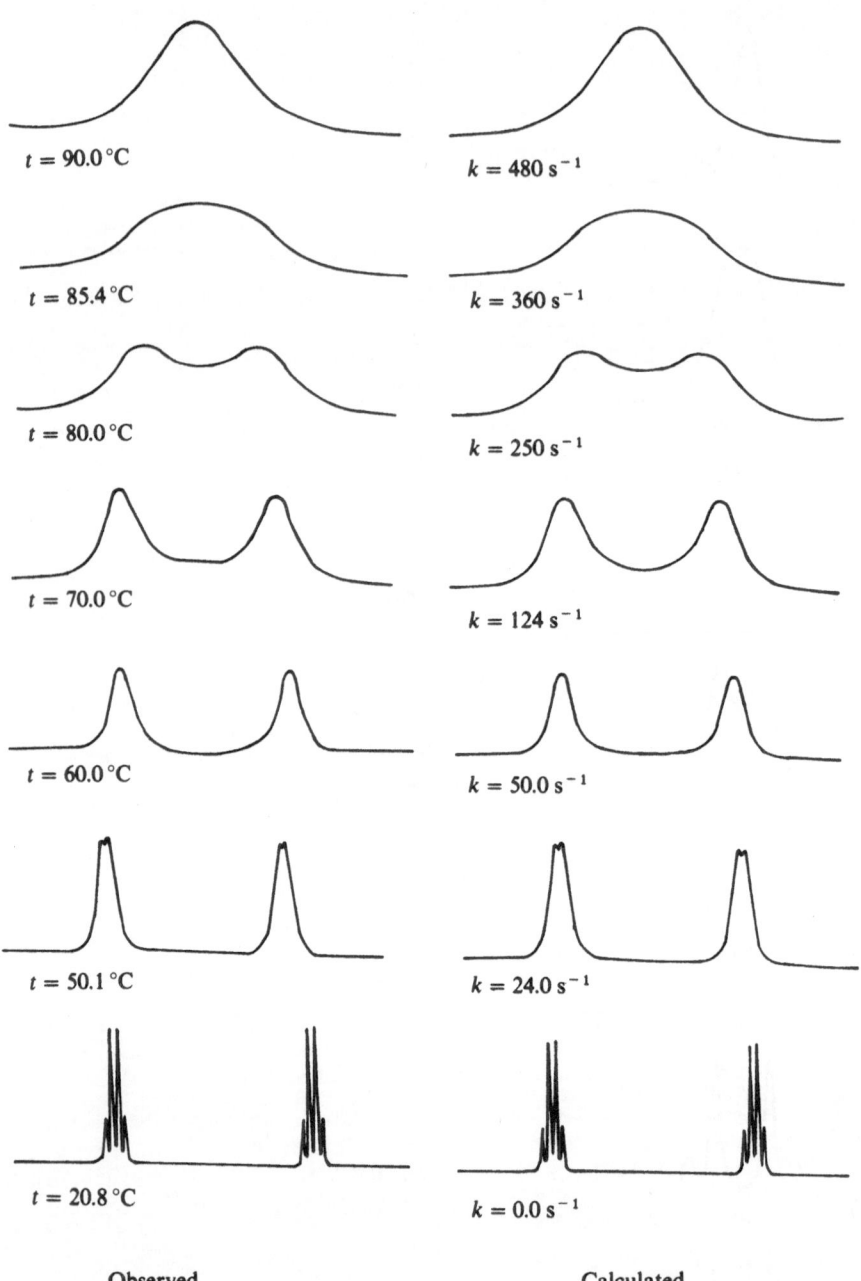

Observed

Calculated

Fig. 3.3. Lineshapes due to methylene protons in N,N-diethylacetamide at various temperatures and calculated spectra with rates of exchange.

time (T_2). Coupling constants do not vary to a significant extent at various temperatures in normal cases but chemical shift differences and T_2's drift as temperature is changed. At temperatures where line shapes change and thus direct reading of the chemical shift differences is not possible, this drift causes difficulty in simulation. Fortunately, however, it is possible to extrapolate the chemical shift difference at coalescence temperature from the known behavior of the differences at low temperatures. That is, the chemical shift differences at the slow exchange limit can be read at several temperatures. They are very often linearly correlated with temperature. Then it is possible to know the difference at a given temperature by linear extrapolation.

The treatment of T_2's is a problem in many line shape analyses, although they are obtained from the line-width at the slow exchange limit or from that of other signals that are not involved in the exchange at a specific temperature. These values do not necessarily give good agreement of the calculated line shapes with those observed. It is a common practise to modify T_2's to obtain the best fit of the calculated spectra with the observed ones.

An example of the simulation, the observed and the calculated line shapes of the methylene protons in N,N-diethylacetamide (**3-3**) is shown in Fig. 3.3. The reader will realize that the agreement of the calculated spectra with those observed is satisfactory.

3-3

3.1.3 Saturation Transfer

When two species are in equilibrium where the rates of exchange are comparable with spin-relaxation time (T_1), then saturation transfer may be used for the determination of the rates of exchange. Suppose signals A and B are due to nuclei which are exchanging their sites. If the signal A is saturated by irradiation, the intensity of the signal B is affected by the irradiation of the signal A. According to the theory, the magnetization of B in the presence of a saturating field at A is a function of T_1 and the rates of exchange. This was first forwarded by Forsén and Hoffman [8]. This method was originally called double resonance because of the technique involved, but since there are many ways of applying the double resonance technique today, the method of obtaining the rates of exchange by double resonance is more appropriately called saturation transfer.

In the days when only CW machines were available, T_1 and the magnetization had to be determined simultaneously. In most cases, it was necessary to observe signals at a rate of once a second or so and if T_1 was short it was necessary to do it even more frequently. This was of course experimentally very difficult and prevented the use of this technique in organic chemistry. However, this difficulty has been removed by the development of modern Fourier transform spectrometers, because they enable us to determine T_1 independently.

In modern FT NMR spectroscopy, the rates of exchange can be measured by either of the following three experiments: a) both signals due to exchanging nuclei are inverted by a nonselective 180° pulse and the recovery of the magnetization is followed with a 90° observation pulse, b) either one of the exchanging signals is inverted by a selective 180° pulse and the return to equilibrium is monitored with a nonselective pulse, and c) either one of the exchanging signals is irradiated with a saturating pulse for a measured time before the observation pulse. The methods were applied to a same process and the results are fully discussed [9].

This method can be applied to the cases where the exchange rates are close to T_1, often meaning that the rates are little slower than those suitable for the line shape analysis. This is convenient in expanding the temperature range where the rates of exchange are obtained, thus making the activation parameters obtained more reliable.

An example is the ring inversion of cis-decalin (**3-4**). This compound possesses five pairs of equivalent carbon: 1 and 5, 2 and 6, 3 and 7, 4 and 8, and 9 and 10. When the ring inversion takes place, C(1,5) and C(4,8) exchange their sites as well as C(2,6) and C(3,7), whereas the C(9,10) pair is unaffected. Although this kinetics was studied by ^{13}C 1D NMR spectroscopy [10], it was also studied by the combination of the line-shape analysis and saturation transfer [11]. By combination of the line shape analysis and the saturation transfer, the following activation parameters for inversion were obtained for a CD_2Cl_2 solution: $\Delta H^{\neq} = (12.35 \pm 0.11)$ kcal/mol, $\Delta S^{\neq}$ (0.15 ± 0.44) e.u.

3-4

3.1.4 The 2-D NMR Spectroscopy

The technique of two-dimensional NMR spectroscopy became available to the problem of stereodynamics in late 1970s. A signal obtained as a function of two independent time variables is Fourier-transformed to afford two-dimensional NMR frequency maps. The signals thus obtained correspond to resonance frequencies of reactants and products in the dynamic process.

This technique is similar in many ways to two-dimensional nuclear Overhauser effect spectroscopy, which is abbreviated to NOESY. However, the cross-relaxation correlations are not desirable in determining the rates of exchange and thus techniques are applied to avoid these cross-relaxation correlations. To avoid the confusion between NOESY, which is valuable in assigning nuclei which are located closely with each other and important in static stereochemistry of complicated molecules, and the technique to obtain the barrier to exchange, the latter is now called EXSY [12].

EXSY is valuable in determining the rates of exchange in complicated systems, which cannot be studied by 1D dynamic NMR, and has been applied to systems which consist of exchange of multi-sites. However, if the signals are strongly coupled, it makes the interpretation of 2D EXSY difficult. Thus, in most of the cases proton-decoupled ^{13}C NMR are used for the EXSY study.

The time scale suitable for this study is longer than that for the line-shape analysis and covers the rates of exchange which are lower than that suitable for the line-shape analysis. Thus this methodology is also important in organic chemistry. However, due to the complexity of the problems, the EXSY technique has not yet become popular among organic chemists. The readers who are interested in the technique are referred to a review [12], which deals with the theoretical as well as the technical aspects to overcome the problems associated with this technique.

The inversion of cis-decalin was studied by this technique and the following activation parameters were obtained: $\Delta H^{\neq} = (13.6 \pm 0.7)$ kcal/mol, $\Delta S^{\neq}$ (3.6 $\pm$ 2.9) e.u. [13]. The results are taken to be favorably compared with those obtained by other methods.

3.1.5 Classical Kinetics

If one can isolate a rotational isomer, then it is possible to follow the rates of isomerization by various means. The reaction must be unimolecular and reversible. Thus the rate constant can be obtained by the usual treatment of the first order and reversible reactions.

If the conversion between rotational isomers is fast, equilibrium is reached rapidly and the population ratio is equal to the equilibrium constant at a given temperature. In these cases, it is necessary to run the kinetics at temperatures, where the rates are suitable to classical methods, and perturb the equilibrium by some means. Here two examples of such perturbation of the equilibrium will be discussed.

The first example is the isomerization of 9-(2-tert-butylphenyl)fluorene (3-5). The ap form of this compound is so unstable with respect to its sp counterpart that the equilibrium constant, sp/ap, is > 100 at 0 °C in CDCl$_3$. The ap form of this compound can be prepared, however, in an almost pure form by acidifying the lithium fluorenide prepared from the compound and butyllithium in ether-THF at − 50 °C, thanks to the solvation of the lithium compound which favors the structure in which the tert-butyl group is over the fluorene ring. Then the isomerization kinetics could be followed by ^{1}H NMR at − 15 to 0 °C [14].

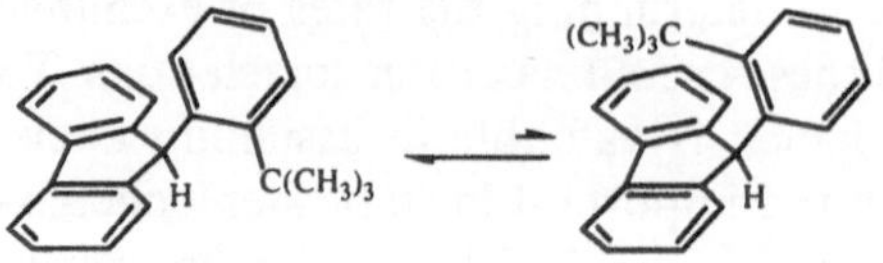

sp-3-5   ap-3-5

Another example is the rapid solvent change to perturb the equilibrium. Equilibrium constants of the rotamers of 9-hydroxy-9-(2-methoxy-1-naphthyl) - 1,4-dimethylfluorene (3–6) favor the sp form in halogenated hydrocarbons due to the presence of the intramolecular hydrogen bond, but favor the ap form in dimethyl sulfoxide or in acetone due to the presence of the strong intermolecular hydrogen bond [15]. The sample of 3-6 was dissolved in CDCl$_3$ and acetone was added to the solution at − 50 °C. The rates of isomerization could be followed between 0 and 11 °C.

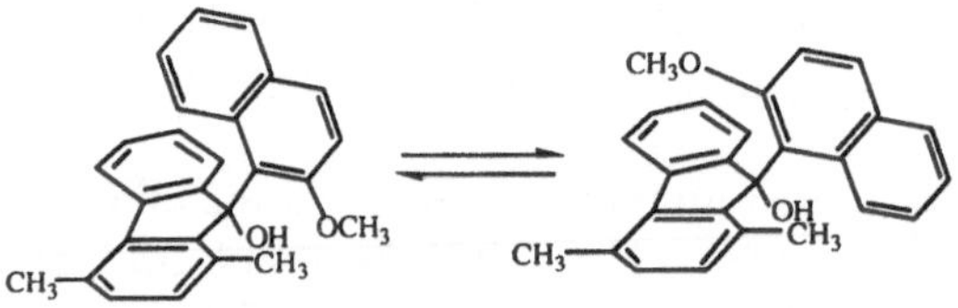

sp-3-6   ap-3-6

3.1.6 Reliability of the Data

So far, various techniques of determining the rate constants of internal rotation have been reviewed without mentioning the reliability of the data thus obtained. How reliable are they?

There is a typical example which deals with rotational barriers in 9-(2-isopropylphenyl)fluorene (3-7) obtained by various methods [16]. The results are shown in Table 3.1 together with the rates of rotation obtained by the classical method, which is believed to afford the most reliable data.

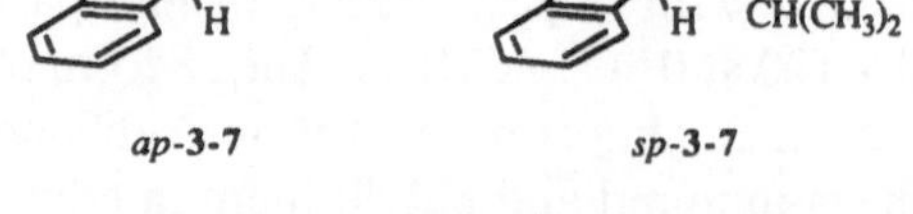

ap-3-7   sp-3-7

As is clear from the data in Table 3.1, ^{1}H dynamic NMR data obtained with the use of isopropyl protons are in excellent agreement with those obtained by the classical method and the reliability is high. Other data obtained by line

Table 3.1. Activation parameters for isomerization of 9-(2-isopropylphenyl)fluorene (3-7) obtained by various methods

Method	$\Delta H^{\neq}$/kcal mol^{-1}	$\Delta S^{\neq}$/e.u.	Percentage error in the slope of the Eyring plot
^{1}H DNMR			
$(CH_3)_2CH$	17.4 ± 0.1	-4.3 ± 0.4	0.65
H-9	17.6 ± 0.4	-3.9 ± 0.4	1.3
^{13}C DNMR			
$(CH_3)_2CH$	17.4 ± 0.9	-4.1 ± 2.3	5.1
C-9	18.0 ± 0.4	-2.2 ± 0.9	1.9
Kinetics of Equilibn.	16.9 ± 0.1	-4.1 ± 0.3	0.52
Saturation Transfer	16.8 ± 0.5	-6.4 ± 1.7	3.2

shape analysis are also in good agreement with others, but close examination discloses that the errors involved are rather large for simple uncoupled signals. The large errors associated with the ^{13}C line shape analysis may also be ascribed to the simple line shapes. Indeed, it is often difficult to determine T_2 and k unequivocally, if the line shapes to be simulated are simple. It can be concluded that line shape analysis gives satisfactory data and the reliability of the data is high if the line shapes are complicated by the presence of couplings.

Saturation transfer also gives satisfactory data as is clear from Table 3.1. However, the reliability of the data is inferior compared to those obtained by the analysis of line shapes of isopropyl protons. This method is probably rated as highly as the line shape analysis of uncoupled nuclei but is inferior to those obtained with use of the line shape analysis of complex signals.

How about the coalescence method? We can now compare the results with those obtained by the total line shape analysis. According to one reference, this coalescence method gives satisfactory data if the chemical shift difference and the coupling constant are in the correct range [17]. The problem here is that, if a chemical shift difference is either too small or too large, reading of the coalescence temperature becomes incorrect. Thus in these cases, one has to admit that a large error is involved.

Because the coalescence method gives rates at one temperature only, it should be used as a rough estimate. Indeed, observation of broadening of the lines in ^{1}H NMR spectra can be taken a sign of slow rotation or other dynamic processes as already pointed out in Chap. 2. Although entropy of activation for internal rotation of molecules is usually very small and free energy of activation by the coalescence method gives a good approximation of the enthalpy of activation, computer simulation of line shapes followed by Arrhenius or Eyring plots affords much more useful information.

3.2 High Barriers to Rotation in the 9-Arylfluorene Series

3.2.1 Isolation of Rotational Isomers

As already pointed out, 9-arylfluorene systems were shown to exhibit a very high barrier to rotation about the C_{aryl}-to-C_9 bond by NMR spectra at high temperatures. The spectra implied that the rotational isomers should be isolable at room temperature. Indeed, the rotational isomers of 9-(2-methyl-l-naphthyl) fluorene (3-8) were partially concentrated and their rotational barrier was determined to be 29.2 kcal/mol at 116 °C [18].

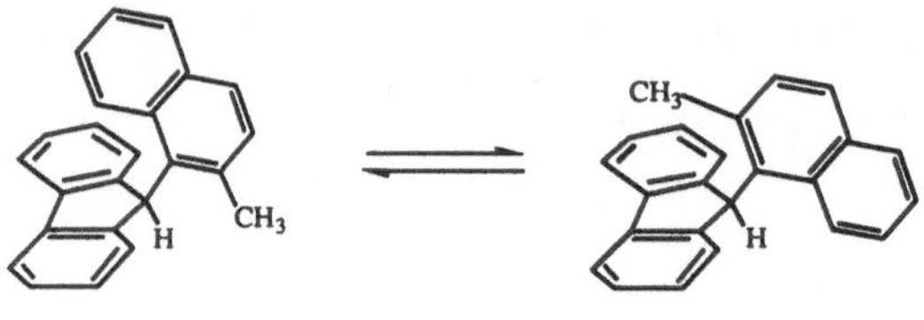

ap-3-8   sp-3-8

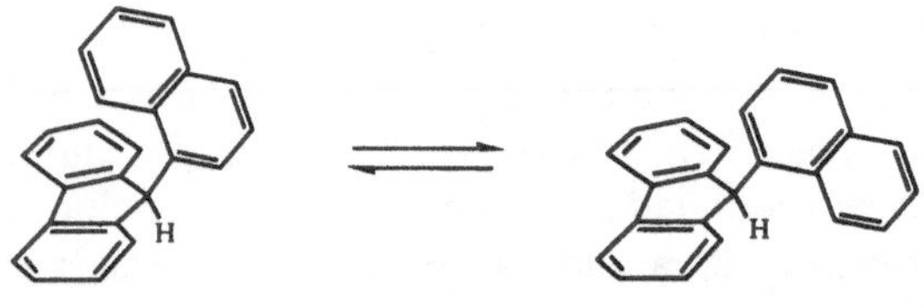

ap-3-9   sp-3-9

However, it is also reported in the same literature that if a methyl group is removed from the compound, the barrier is lowered considerably, namely to 18.0 kcal/mol at 60 °C, and it is now impossible to isolate stable rotational isomers of this compound (3-9) at room temperature. Therefore, it seems very important for an aryl group attached to the 9-position of the fluorene ring to possess two substituents at the o-positions, in order to make it possible to isolate the rotational isomers in this system. Then it is interesting to see the effect of the substituents on the rotational barrier in this system. The point will be further elaborated in Sect. 3.2.2.

The barrier for the dynamic process in these compounds, it may be argued, is too high for rotation: Is it not bond breaking followed by recombination but really an internal rotation that we observe? The answer to this question was elegantly provided by Ford et al. in the following way [19].

Compounds designed for this purpose are 9-substituted 3-isopropyl-9-mesitylfluorenes (3-10). The C-9 of the fluorene ring in these compounds is a chiral center. Thus the isopropyl-methyl protons in these compounds are diastereotopic. If either the C_9-to-mesityl bond or the C–X bond is broken during the process we are observing, it is natural to expect that the magnetic sites of the isopropyl-methyl protons should be exchanged with concomitant exchange of the *o*-methyl protons of the mesityl group. On the other hand, if it is a pure bond rotation, the proton signals due to the *o*-methyls in the mesityl group should exchange their sites, whereas the isopropyl signals remain intact. Experimentally, the latter is the case, thus proving that the dynamic process is indeed internal rotation.

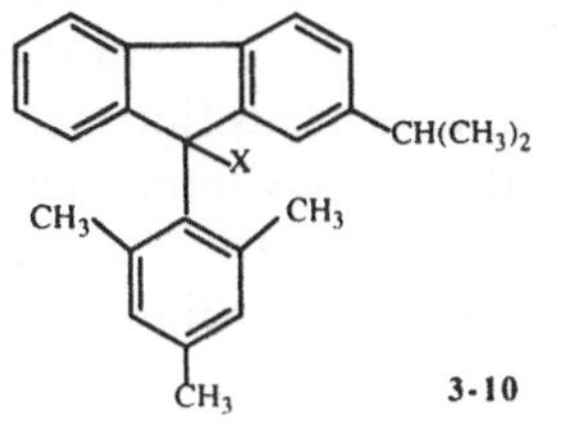

3-10

Since 9-mesitylfluorene (**3-11**) was shown to possess a higher barrier to rotation than 25 kcal/mol by the dynamic NMR method, it will be a natural extension to modify one of the *o*-methyl groups for isolation of rotational isomers. This was accomplished by introducing a bromo substituent to one of the methyl groups of 9-(2,6-dimethylphenyl)fluorene (**3-12**) [20].

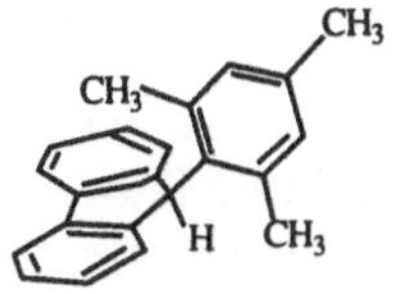

3-11 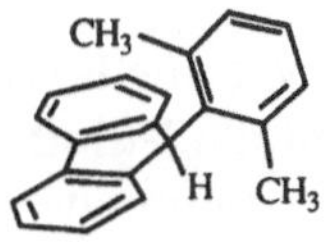   3-12

The rotational isomers of 9-(2-bromomethyl-6-methylphenyl)fluorene (**3-13**) were thus isolated and the rotational barrier was determined to be ca. 29 kcal/mol. This barrier is about the same with that in 9-(2-methyl-l-naphthyl) fluorene and indicates that a methyl group is as effective as a *peri*-CH group in the naphthalene nucleus in determining a high barrier to rotation.

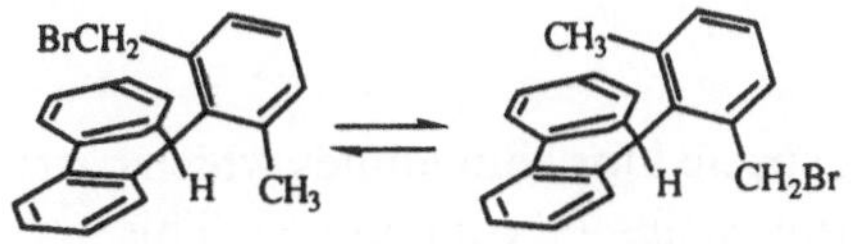

ap-3-13   *sp*-3-13

3.2.2 Factors Affecting the Rotational Barriers

3.2.2.1 Effects of the Steric Size of the Substituents

It may be necessary to see the effect of the bromo-substituent in the compound on the rotational barrier, if one really wants to say that the effects of the methyl group and the *peri*-CH of the naphthalene are almost the same from the data on compounds **3-8** and **3-13**. This was done by synthesizing variously substituted compounds at the methyl group of 9-(2-methyl-1-naphthyl)fluorene (**3-14**) [21]. The barrier was practically unaffected by the substituent. Whereas the barrier to rotation in the parent compound is 29.0 kcal/mol at 373 K, those of the compounds carrying a bromo, hydroxy, and methoxy groups are 29.9, 29.0, and 29.3 kcal/mol, respectively. This may be ascribed to the fact that the substituent can rotate to take a conformation in which the substituent is away from the interacting site so that it does not affect the ground state and the transition state of rotation. Strictly speaking, this situation should affect the entropy terms. The present results imply that the entropy term is not important in these cases, probably because the change in the ground state and that in the transition state are similar.

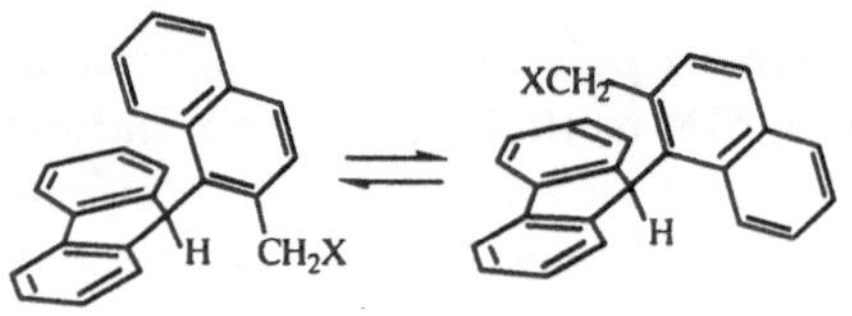

sp-**3-14**   ap-**3-14**

When the XCH_2 group is replaced by a methoxy group, the barrier to rotation is reduced to a considerable extent. 9-(2-Methoxy-1-naphthyl)fluorene (**3-15**) and 9-(2-methoxy-4,6-dimethylphenyl)fluorene (**3-16**) give barriers to rotation from the *ap* isomer to the *sp* of 24.1 and 26.3 kcal/mol respectively, a decrease of ca. 4 kcal/mol. This is ascribed to the small size of the methoxy group compared to the substituted methyl group [22].

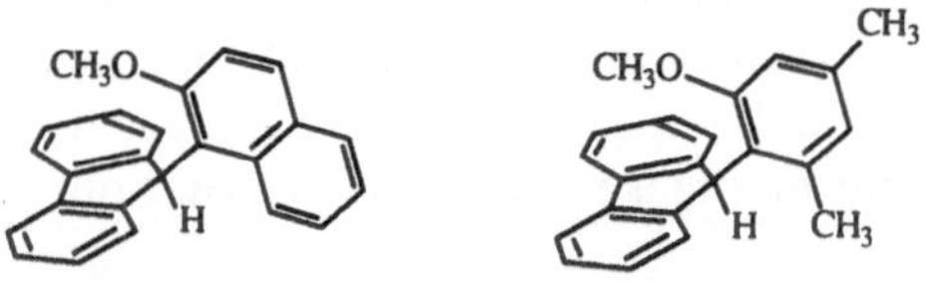

ap-**3-15**   ap-**3-16**

A similar effect due to the steric size is expected for compounds which carry an amino group in the 9-aryl group of this series of compounds. This was examined by taking 9-(2-amino-1-naphthyl)fluorene derivatives (**3-17**). The barriers to rotation of 9-(2-amino-1-naphthyl)fluorene, 2-methylamino, and 2-dimethylamino compounds for the process, *sp* → *ap*, were 27.9, 27.8, and

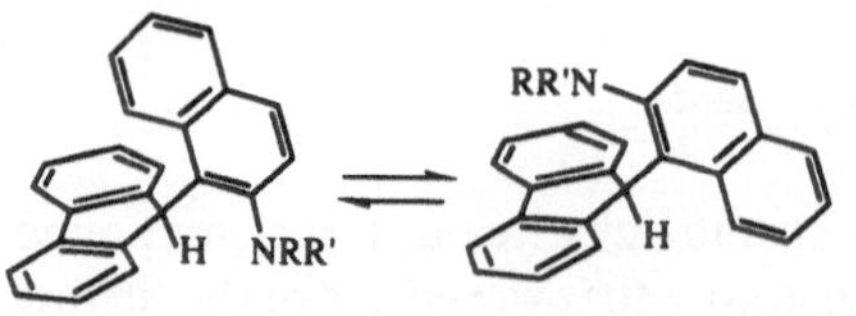

sp-**3-17**   ap-**3-17**

27.4 kcal/mol at 80.4 °C, respectively [23]. As expected on steric grounds, these barriers are a little higher than those of the corresponding methoxy compounds. However, the barriers are not affected by the methyl substitution on the amino-nitrogen. This shows that the second atom of the 2-substituent in the series of compounds does not affect the barrier. However, the barrier to rotation for the reverse process, $ap \rightarrow sp$, changes according to the substitution patterns. They are 29.0, 28.9 and 26.1 kcal/mol for the amino, methylamino, and dimethyl-amino compounds, respectively. This reflects the stabilization of the ground state, as will be discussed later in this section, due to the NH-π bond, though it is weak.

The simplest example of reducing the barrier to rotation in the 9-arylfluor-ene series is 9-(2-alkylphenyl)fluorenes (3-18) which lack a methyl group in one of the flanking o-positions of 9-(2,6-dimethylphenyl)fluorene skeleton. These are also analogs of 9-(1-naphthyl)fluorene. The barriers to rotation from the ap form to the sp and vice versa are listed in Table 3.2 [14]. Apparently, the size of the alkyl group at the o-position of the phenyl group affects the barrier to indicate that, at the transition state for rotation, the interaction of the o-substituent and the $peri$-CH of the fluorene ring is important and the phenyl group is rotating without significant deformation. The most noteworthy fact is, however, that the o-alkyl group affects the ground state. Whereas other compounds do show the presence of the ap-isomer, 9-(2-$tert$-butylphenyl)fluorene (3-18: R=(CH$_3$)$_3$C) exists solely as the sp isomer due to the steric effect caused by the bulkiness of the $tert$-butyl group which interacts with the fluorene ring. Thus it is not possible to obtain the barrier to rotation from the sp form to the ap for the $tert$-butyl compound.

ap-3-18   sp-3-18

Table 3.2. Activation parameters for rotation in 9-(2-alkylphenyl)fluorene (3-18)

R	Process	ΔH^*/kcal mol^{-1}	ΔS^*/e.u.	$\Delta G^{\ddagger}_{273}$/kcal mol^{-1}
H	$sp \rightarrow ap$	—	—	< 9
	$ap \rightarrow sp$	—	—	< 9
CH$_3$	$sp \rightarrow ap$	15.9	− 2.4	16.6
	$ap \rightarrow sp$	15.9	− 1.5	16.3
C$_2$H$_5$	$sp \rightarrow ap$	18.2	1.1	17.9
	$ap \rightarrow sp$	18.2	3.3	17.3
(CH$_3$)$_2$CH	$sp \rightarrow ap$	16.0	− 8.6	18.3
	$ap \rightarrow sp$	16.0	− 7.3	18.0
(CH$_3$)$_3$C	$sp \rightarrow ap$	—	—	> 23
	$ap \rightarrow sp$	21.8	5.6	20.3

Buttressing effects are well documented in the barrier to rotation in the biphenyl series [24]. The effects are present because, in the transition state of rotation, a substituent, which otherwise either takes on other conformations or escapes the steric repulsion by bond-angle deformation, cannot take on such conformation or deformation due to the presence of a neighboring substituent. In the famous biphenyl series, the effects are thought to be especially strong because both conformational change and the angle deformation are prevented by the neighboring substituent.

In this context, it is interesting to examine the barriers to rotation in 9-(2-methoxy-4,6-dimethylphenyl)fluorene derivatives (3-19), because the neighboring substituent to the 2-methoxy group will prevent the methoxy group taking the conformation away from the fluorene moiety in the transition state of rotation. The results of introducing bromo substituent(s) to the mother compound are listed in Table 3.3 [25].

Clearly, the bromo substituent introduced at the 3-position of the phenyl group has a stronger enhancing effect on the barrier to rotation than that in the 5-position which is next to the methyl group. This effect must be caused by the fact that the methoxy-methyl group cannot take a position away from the interacting site with the fluorene ring due to the presence of the buttressing substituent. The importance of the conformational effect in the buttressing effect is thus proved.

ap-3-19   sp-3-19

When one is considering the energy of the transition state for rotation, the above discussion is correct. However, as pointed out earlier, the barrier height is the difference between the ground state and the transition state. If the ground state energy is increased more than the increase in the transition state energy, the observed barrier height will be diminished. An example of this category is found in 1-substituted 9-arylfluorene derivatives (3-20).

A simple consideration of the transition state conformation of 9-arylfluorene derivatives suggests that both interactions, 1-CH and 8-CH of the fluorene ring,

Table 3.3. Activation parameters for rotation ($sp \rightarrow ap$) in compound 3-19 in tetrachloroethene

X	Y	$\Delta H^{\neq}$/kcal mol^{-1}	$\Delta S^{\neq}$/e.u.	$\Delta G^{\neq}_{329}$/kcal mol^{-1}
H	H	24.4 ± 0.1	− 1.9 ± 0.1	25.0
H	Br	22.2 ± 0.3	− 11.3 ± 0.8	25.9
Br	H	23.2 ± 0.0	− 10.0 ± 0.0	26.5
Br	Br	25.4 ± 0.1	− 5.4 ± 0.4	27.2

with the 9-aryl group are important in determining the barrier height for rotation. Therefore, if one introduces a substituent to either 1 or 8 position of the fluorene moiety, it should enhance the barrier. The effect of the 1-substituent on the rotational barrier in 9-(2-methyl-1-naphthyl)fluorene is summarized in Table 3.4 [26], in which the process shown in the following scheme is taken. Because of the complexity of nomenclature of stereochemistry, its designation by symbols is not given.

The methyl group enhances the barrier as expected and the halogen substituent does so as the size of the substituent increases. However, when a fluorine substituent is introduced, the barrier height is reduced relative to the compound without the substituent.

The results must be attributed to the increase in the ground state energy of the fluoro compound due to the steric effect. To acquire information about the deformation of the molecule, X-ray crystallography was carried out with the crystals of the fluoro compound and the related compounds. The 1-fluoro compound showed difference in several bond lengths and bond angles, when compared with those of the unsubstituted compound, to accommodate the steric strain caused by the substitution. This destabilization of the ground state must be responsible to lowering the barrier to rotation compared with the unsubstituted compound. The other substituents should also give steric strain but their effects of enhancing the transition state energy seem to outweigh the destabilization of the ground state.

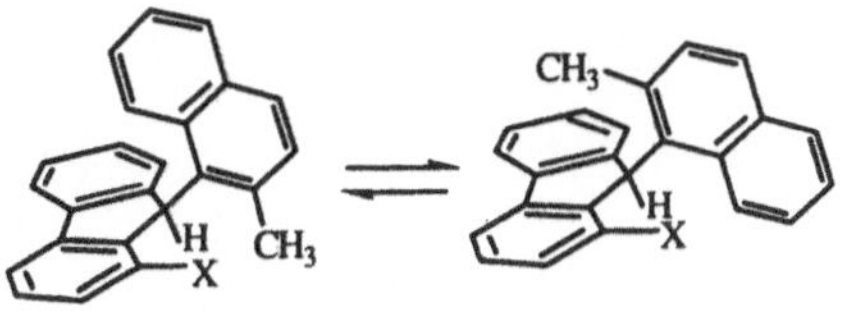

3-20

It seems, however, the 1-methyl group in the fluorene moiety effectively enhances the barrier to rotation, in the cases where the compound carries a 9-OH group. The compound (**3-21**) was reported to show the barrier to rotation of more than 26 kcal/mol [19] and one of the rotational isomers of a benzo-annelated derivative (**3-22**) was isolated as a stable compound at room temperature. The barrier to rotation was 24.6 kcal/mol [27].

Table 3.4. Activation parameters and relative rates on internal rotation in compound **3-20** in o-dichlorobenzene

X	$\Delta H^{\neq}$/kcal mol^{-1}	$\Delta S^{\neq}$/e.u.	$\Delta G^{\neq}_{383}$/kcal mol^{-1}	k_{rel}(383 K)
H	25.5 ± 0.2	-11.2 ± 0.6	29.7	1.0
F	24.8 ± 0.5	-11.6 ± 1.5	29.3	1.8
Cl	27.9 ± 0.8	-10.3 ± 0.8	31.8	0.067
Br	28.1 ± 0.9	-10.8 ± 0.9	32.2	0.038
CH$_3$	28.2 ± 1.5	-11.7 ± 3.8	32.7	0.019

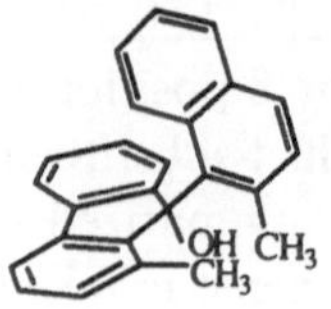

3-21 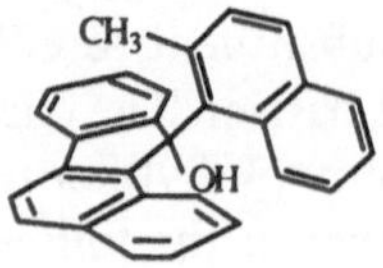   3-22

Compounds (**3-23**) which bear an 8-methyl group in the naphthalene ring of 9-(1-naphthyl)fluorene were also prepared to find that the barrier to isomerization from the form, in which the 8-methyl group is up on the fluorene ring, to another, in which the 8-methyl group is in nearly the plane of the fluorene, was enhanced to 23.9 kcal/mol at 307 K, when X is H [28]. This must mean that in the transition state for rotation the interaction between the 8-methyl group and the 1 and/or 8-CH groups of the fluorene ring is severe, because the apparent barrier to rotation is enhanced with respect to the compound which lacks the 8-methyl group, although the ground state is destabilized by the strong interaction between the 8-CH$_3$ group and the fluorene ring. Interestingly, introduction of a methyl group at the 1-position of the fluorene ring enhanced the barrier to rotation to 25.2 kcal/mol at 307 K. As expected, the equilibrium constant is in favor of the form in which the 8-methyl group is in nearly the plane of the fluorene ring.

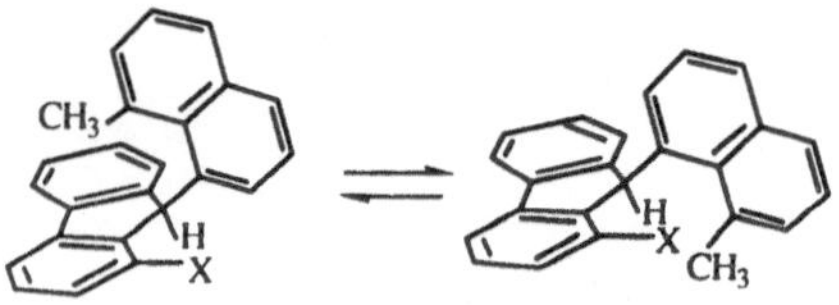

3-23

3.2.2.2 Ground State Destabilization

Clearer examples of lowering the barrier to rotation by destabilizing the ground state structure are found in the series of 9-substituted 9-arylfluorenes. The first example was provided by Chandross and Sheley [29] by showing that the barrier to rotation is the lowest when the 9-substituent is chlorine and it is the highest when there is no substituent at the 9-position in 9-mesitylfluorenes (**3-24**). The ^{1}H NMR spectrum of the chloro compound gives a broad singlet for

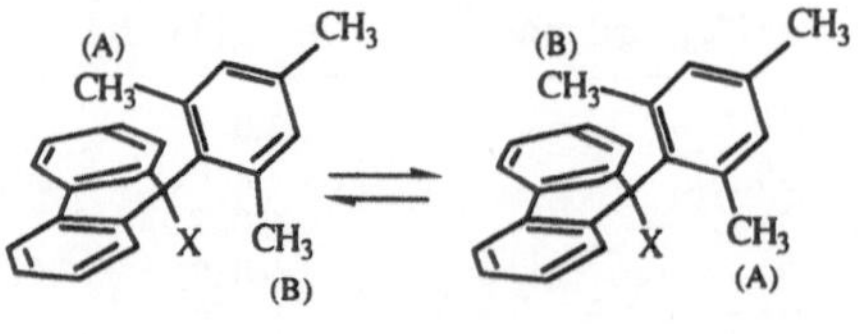

3-24

the two methyl groups in the mesityl moiety at room temperature, whereas that due to the mother compound is a distinct doublet, which does not coalesce even at 200 °C.

This effect was also studied by other authors for 9-substituted 9-(2,6-dimethoxyphenyl)fluorenes (3-25). The results are summarized in Table 3.5 [30]. Clearly, the introduction of a bulky substituent in the 9-position of the fluorene system reduces the barrier to rotation irrespective of the substituent in the aryl group of the 9-arylfluorene system. Not only the compounds listed in Table 3.5 but also other compounds of the similar type are known to show lowering of the barrier to rotation on introduction of substituents to the 9-position [31]. The increase in the ground state energy is important in these systems in lowering the rotational barrier.

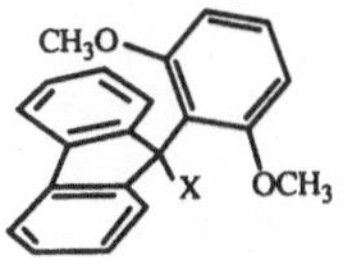

3-25

Rotational barriers for the 2-substituted 1-(9-fluorenyl)naphthalenes (3-26) in which the substituent involves a double bond are understood as examples of destabilization of the ground state as well. The results of the cases where the double bond is a part of olefins are shown in Table 3.6 [32, 33].

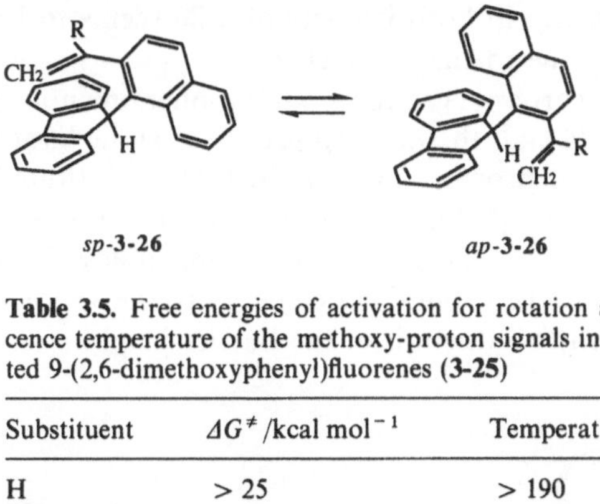

sp-3-26   ap-3-26

Table 3.5. Free energies of activation for rotation and coalescence temperature of the methoxy-proton signals in 9-substituted 9-(2,6-dimethoxyphenyl)fluorenes (3-25)

Substituent	$\Delta G^{\neq}$/kcal mol^{-1}	Temperature/°C
H	> 25	> 190
OH	20.2	145
Cl	16.2	66

Table 3.6. Free energies of activation for rotation of 9-arylfluorenes carrying an olefinic group (3-26) in benzene-d_6

R	Process	$\Delta G^{\neq}$/kcal mol^{-1}	Temperature/°C
H	$sp \rightarrow ap$	28.4	80
	$ap \rightarrow sp$	28.9	80
CH_3	$sp \rightarrow ap$	27.5	80.5
	$ap \rightarrow sp$	29.4	80.5

Table 3.7. Free energies of activation for rotation in 9-[(2-substituted carbonyl)-1-naphthyl]fluorene (**3-27**) at 55°C in benzene-d_6

X	Process	ΔG^*/kcal mol^{-1}
H	$sp \rightarrow ap$	26.9
	$ap \rightarrow sp$	26.7
CH$_3$	$sp \rightarrow ap$	26.3
	$ap \rightarrow sp$	24.6
C$_6$H$_5$	$sp \rightarrow ap$	26.6
	$ap \rightarrow sp$	24.9
OH[a)	$sp \rightarrow ap$	25.7
	$ap \rightarrow sp$	23.8
OCH$_3$	$sp \rightarrow ap$	26.6
	$ap \rightarrow sp$	25.3

[a) DMSO solvent.

When R is hydrogen, the rate constants of isomerization, $sp \rightarrow ap$ and $ap \rightarrow sp$, are comparable. However, when R is a methyl group, the rate constant for isomerization from the sp form to the ap is much higher than that for the reverse step. This is because the ap form is unstable with respect to the sp due to the fact that the l-methylvinyl group in that form interacts very strongly with the fluorene moiety. The l-methylvinyl group in both isomers of **3-26** seems to be nonplanar with the naphthalene ring due to the steric effects.

When the double bond is a part of carbonyl groups, a similar phenomenon is observed. As is shown in Table 3.7 [34], all the $ap \rightarrow sp$ processes show larger rates of isomerization than the reverse process, $sp \rightarrow ap$, in 9-[(2-substituted carbonyl)-1-naphthyl]fluorenes (**3-27**) except for the formyl case (X=H). The carbonyl groups cannot be coplanar with the naphthalene ring unless X is hydrogen in the ap forms. Thus, raising the ground state energy of the ap form with respect to the sp form results in the reduction of the barrier height.

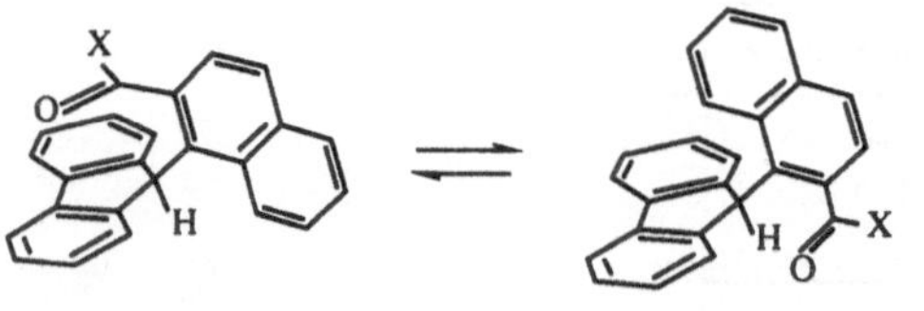

ap-**3-27**   sp-**3-27**

3.2.2.3 Ground State Stabilization

So far, it has been said that an increase in the ground state energy could result in a lowering of the rotational barrier. As a corollary, it should be expected that, if one could stabilize the ground state without affecting the transition state energy, enhancement of the rotational barrier should be observed. One such example is known in the case of the formation of the hydrogen bond.

Hydrogen bond formation between the hydroxyl group and the π-system of the fluorene ring in 9-(2-hydroxy-l-naphthyl)fluorene (**3-28**: R=H) and 9-(2-hydroxy-4,6-dimethylphenyl)fluorene (**3-29**: R=H) is confirmed, because the population of the *ap* rotamers in these compounds are unusually rich, when one compares the data with the corresponding methoxy compounds (**3-28** and **3-29**: R=CH$_3$), and the infrared spectrum is in agreement with this conclusion [22]. The presence of OH-π interactions is well documented and the proximity of the hydroxyl group in the *ap* form to the fluorene moiety should permit the two groups concerned to interact with each other.

sp-**3-28**   ap-**3-28**

sp-**3-29**   ap-**3-29**

In contrast, in the transition state for rotation, the main interaction will be that between the 1-CH (8-CH) group of the fluorene moiety and the *o*-substituent (or 8-CH of the naphthyl group). Here the interaction between the oxygen of the hydroxy group or the methoxyl group is important. However, if a molecule rotates internally with a minimum energy required, in the transition state, the methyl group in the methoxy will point away from the interacting site of the fluorene system. Thus it is expected that the transition state energy for rotation in the methoxy compound will not differ much from that in the hydroxy compound.

The results are in accord with the expectation as is shown in Table 3.8 [22]. The hydroxy compounds give a ca. 1.0 kcal/mol higher barrier to rotation in the *ap* → *sp* process, whereas in the *sp* → *ap* process the barriers are almost the same. This energy for the OH-π interaction is in the range of the normal value for such an interaction.

It is well known that the N–H group forms a weaker hydrogen bond than an O–H group if an electron donor is fixed: the NH–O hydrogen bond is weaker than the OH–O bond. It is now known that the N–H stretching frequency is not affected by the formation of the NH-π bond [23]. In this context, compounds that possess an amino group which is capable of forming a hydrogen bond with the π-system of the fluorene moiety should show a less distinct effect on the

Table 3.8. Kinetic parameters for rotation in 9-arylfluorenes carrying an oxygen-substituent (**3-28** and **3-29**)

Compound	Process	$\Delta H^{\neq}$/kcal mol^{-1}	$\Delta S^{\neq}$/e.u.	$\Delta G_{329}^{\neq}$/kcal mol^{-1}
3-28 (R=H)	$sp \rightarrow ap$	22.2 ± 0.2	-10.9 ± 0.4	25.7 ± 0.1
	$ap \rightarrow sp$	23.0 ± 0.3	-10.1 ± 0.9	26.3 ± 0.1
3-28 (R=CH$_3$)	$sp \rightarrow ap$	24.2 ± 0.2	-5.2 ± 0.7	25.9 ± 0.1
	$ap \rightarrow sp$	22.7 ± 0.2	-7.3 ± 0.6	25.1 ± 0.1
3-29 (R=H)	$sp \rightarrow ap$	22.4 ± 0.1	-7.1 ± 0.3	24.7 ± 0.1
	$ap \rightarrow sp$	23.0 ± 0.1	-6.5 ± 0.2	25.1 ± 0.1
3-29 (R=CH$_3$)	$sp \rightarrow ap$	24.5 ± 0.1	-1.9 ± 0.1	25.0 ± 0.1
	$ap \rightarrow sp$	23.0 ± 0.1	-3.5 ± 0.1	24.1 ± 0.1

barrier. The results [23] indicate that, whereas the barrier to rotation in the $sp \rightarrow ap$ process is almost the same for the compounds (ca. 27.5 kcal/mol at 80 °C) irrespective of the number of methyl groups on the amino-nitrogen of 9-(2-amino-l-naphthyl)fluorene (**3-17**), it is ca. 26 kcal/mol when all the hydrogens on the amino group are replaced by methyl groups and it is ca. 29 kcal/mol when at least one hydrogen remains on that amino-nitrogen in the $ap \rightarrow sp$ process.

sp-**3-17**   ap-**3-17**

The S–H group, being more weakly acidic than an N–H group in hydrogen bond formation, is expected to show an effect of the hydrogen bond (SH-π) on the rotational barrier too small to observe. In accordance with this expectation, the barriers to rotation of the $ap \rightarrow sp$ and $sp \rightarrow ap$ processes in l-(9-fluorenyl)-2-naphthalenethiol (**3-30**) are 30.2 and 31.2 kcal/mol, respectively [23]. Now, the barrier to the first process is lower than that for the second process. The barriers are very close to those in the corresponding methyl ethers, where they are 29.6 and 31.0 kcal/mol, respectively, for the $ap \rightarrow sp$ and for the $sp \rightarrow ap$ processes [35].

sp-**3-30**   ap-**3-30**

Table 3.9. Free energies of activation for rotation in 1-(9-fluorenyl)-2-naphthyl aroates (**3-31**) in CDCl$_3$ at 69 °C

X	$\Delta G^{\neq}$/kcal mol^{-1}	
	$sp \rightarrow ap$	$ap \rightarrow sp$
CH$_3$O	27.3 ± 0.1	27.2 ± 0.1
CH$_3$	27.3 ± 0.1	27.2 ± 0.1
H	27.3 ± 0.1	27.3 ± 0.1
NO$_2$	27.3 ± 0.1	28.1 ± 0.1

An example of stabilization by charge-transfer interactions is given by a series of 1-(9-fluorenyl)-2-naphthyl aroates (**3-31**). The rate constants for the isomerization are shown in Table 3.9 [36]. As is clear from the table, the rates of isomerization from the *sp* form to the *ap* is not affected by the 4-substituent in the aroyl group. This result supports the idea that the transition energy for the isomerization is also about the same as that expected from the consideration of the transition structure. However, the rate constants for the isomerization from the *ap* to the *sp* form become small as the 4-substituent becomes electron-withdrawing. This tendency is clearly seen, if one compares the results of the nitro compound with others. The results must be derived from the fact that the *ap* form of the nitro compound is specifically stabilized by the charge-transfer interaction between the *p*-nitrobenzoyl group and the fluorene ring as was discussed earlier.

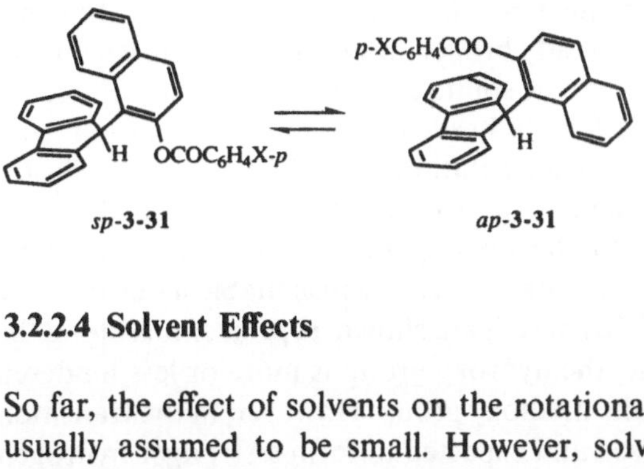

sp-**3-31**   *ap*-**3-31**

3.2.2.4 Solvent Effects

So far, the effect of solvents on the rotational barrier has been neglected. It is usually assumed to be small. However, solvents can affect, in principle, the barrier heights. Indeed, the difference between the 2-formyl (**3-27**: X=H) and 2-vinyl (**3-26**: R=H) compounds of the above 2-substituted 1-(9-fluorenyl)-naphthalene could be attributed to the difference in the solvent used for the determination of the barrier.

If a solvent stabilizes the ground state for rotation but does not affect the transition state, then the barrier should be enhanced. And if a solvent stabilizes the transition state without affecting the ground state, the barrier height is reduced. A very large solvent effect of this type is known for a 9-arylfluorene derivative.

Table 3.10. Solvent effects on activation parameters for rotation in 9-(2-methoxy-1-naphthyl)-fluoren-9-ol (**3-32**) at 30 °C

Solvent	Process	ΔH^{*} /kcal mol^{-1}	ΔS^{*} /e.u.	ΔG^{*} /kcal mol^{-1}
HCB[a]	$sp \rightarrow ap$	16.5	− 3.2	17.5
	$ap \rightarrow sp$	15.2	− 6.6	17.2
DMSO	$sp \rightarrow ap$	12.2	− 10.0	15.4
	$ap \rightarrow sp$	14.1	− 8.6	16.7

[a] Hexachlorobutadiene.

The compound in question is 9-(2-methoxy-1-naphthyl)fluoren-9-ol (**3-32**). The kinetic parameters for rotation in this compound in hexachlorobutadiene and in dimethyl sulfoxide are given in Table 3.10 [37]. The results indicate that the internal rotation is easier in dimethyl sulfoxide than in hexachlorobutadiene. It is especially distinct for the process $sp \rightarrow ap$.

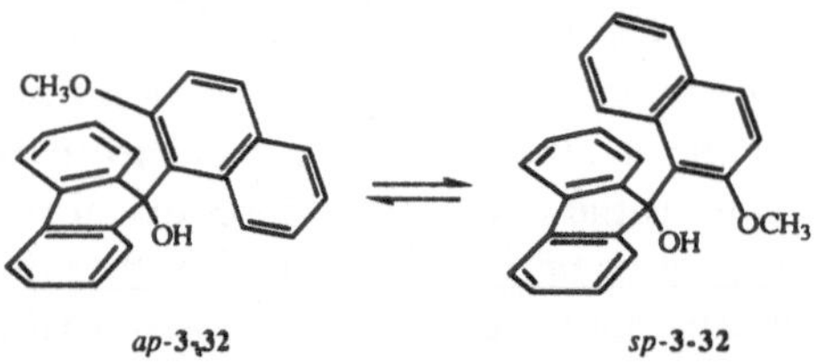

ap-**3-32**   sp-**3-32**

In hexachlorobutadiene, the compound will exist as a strongly hydrogen-bonded form between the 9-OH group and the methoxy group in the sp-form. For rotation, this hydrogen-bond must be cleaved. By contrast, in dimethyl sulfoxide, the hydrogen bonding is already broken in the ground state because of the strong hydrogen-bonding between the solute and the solvent. The difference in the kinetic parameters obtained for the hexachlorobutadiene solution from those obtained for the dimethyl sulfoxide solution is attributed to this effect.

The reduction in the barrier height for the process $sp \rightarrow ap$ in dimethyl sulfoxide, therefore, with respect to the $ap \rightarrow sp$ process must be caused by stabilization of the transition state for rotation. This is reasonable if one looks at the following molecular models (**3-33**), which are shown as projections along the fluorene plane. In the ground state, the hydroxy group is more or less hindered by the naphthalene nucleus or by the methoxy group. However, in the transition state, the OH group is more exposed and expected to form a stronger hydrogen

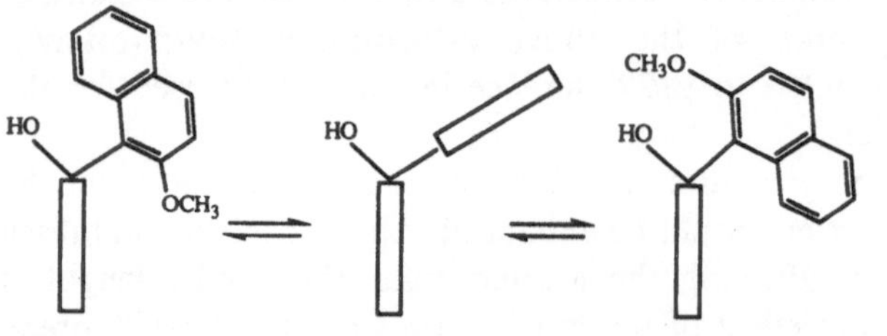

3-33

bond than in the ground state. The difference in the strength of the hydrogen bond must be responsible for the observed activation parameters. In accordance with this interpretation, the entropy of activation for the internal rotation in dimethyl sulfoxide is large and negative for both processes with respect to those in hexachlorobutadiene.

3.3 High Barriers to Rotation in 9-Substituted Triptycene Series

Since the isolation of rotational isomers of 9-(2-phenyl-1,1-dimethylethyl)-triptycene series, a number of factors that affect their rotational barrier have been studied. As expected, the classes of the 9-alkyl substituent affect the barrier to a large extent as well as the substituent at the 1-position. This section describes the effects of the substituent on the rotational barrier.

3.3.1 Triptycenes Carrying a 9-*tert*-Alkyl Substituent

The rotational barriers in substituted 9-(2-phenyl-1,1-dimethylethyl)triptycenes (**3-34**) are given in Table 3.11 [38]. There is an interesting trend in the barrier: the barrier is high if the substituent at the 1-position is rather small and is reduced as the bulkiness of the 1-substituent increases. Close examination discloses that the barrier is at a maximum when the 1-substituent is a fluoro group.

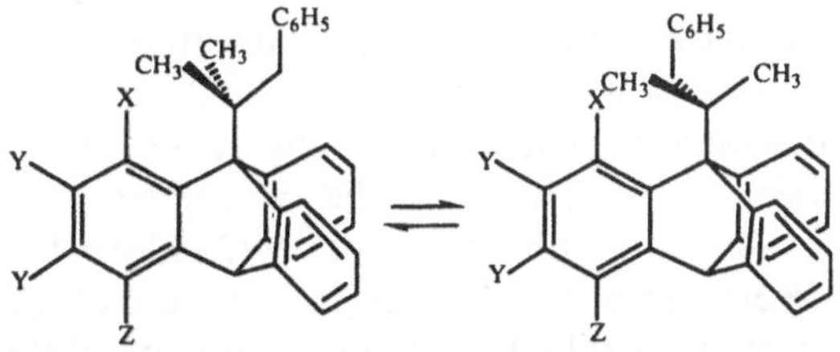

ap-**3-34**   sc-**3-34**

Table 3.11. Free energy of activation for and relative rates of rotation in substituted 9-(1,1-dimethyl-2-phenylethyl)triptycene (**3-34**) for $ap \rightarrow sp$ process at 500 K in 1-chloronaphthalene

X	Y	Z	$\Delta G^{\neq}$ /kcal mol^{-1}	Relative Rates
H	Cl	H	40.4	20
F	H	H	43.2	1
CH_3O	H	CH_3O	42.5	2
Cl	H	H	40.4	18
Cl	H	Cl	39.9	30
Br	H	H	39.2	60
CH_3	H	CH_3	38.4	150
CF_3	H	H	33.8	16000

An interpretation of this interesting phenomenon is that there is a more effective destabilization of the ground state for rotation than the transition state, when the 1-substituent is bulky. If a substituent is small, the ground state is more relaxed but the transition state is also stabilized with respect to compounds with a large 1-substituent. Thus a maximum is observed, when a series of substituents is examined, at an intermediate substituent size. This phenomenon is a kind of buttressing effect but the effect decreases the barrier when the substituent is large, rather than increasing it as is the case in the normal buttressing effect. For the cases in which the buttressing effect reduces the barrier height, the term "negative buttressing effect" is proposed.

ap-**3-35**   *sc*-**3-35**

It is still possible that the substituent at the 2-position exerts buttressing effect on the 1-substituent in the observed barriers in Table 3.11. Thus barriers in compounds (**3-35**) which carry four substituents at the 1,2,3,4-positions were examined. The results are summarized in Table 3.12 [39]. Clearly the 2-substituents give buttressing effects. If a substituent is small, the buttressing effect is positive: the presence of the 2-substituent enhances the barrier. However, if the 1-substituent is large, the buttressing effect is again negative.

The substituent introduced in place of a hydrogen atom in the 9-*tert*-butyl seems not to exert effects on the rotational barrier of the 9-substituent. This is clear, if one compares the barrier to rotation of the 9-(2-phenyl-1,1-dimethylethyl) compound (**3-35**: X=Cl) with that of 9-(2-chloro-1,1-dimethylethyl) compound (**3-36**). Both gives the barrier to rotation of ca. 38 kcal/mol at 500 K. The result is rationalized by knowing the structure of the compounds. Both spectral [40] and X-ray crystallographic [41] evidence indicates that the substituent on the *tert*-butyl group is upright to the

Table 3.12. Free energy of activation for rotation in 1,2,3,4-tetrahalo-9-(1,1-dimethyl-2-phenylethyl)triptycene (**3-35**) and relative rates of rotation to those in compound **3-34**

X	Temp/°C	$\Delta G^{\neq}$/kcal mol^{-1}	k_4/k_1[a]
F	240	44.0	0.5
Cl	175	37.8	13
Br	175	34.7	100

[a] k_4 is the rate constant for the tetrahalo compound and k_1 for the 1-halo compound.

triptycene skeleton, due to the steric effects. Therefore, the substituent affects neither the ground state energy nor the transition state energy. Even 9-(2-chloro-1-chloromethyl-1-methylethyl) compound (**3-37**) gives the same barrier height for rotation with compound **3-36**.

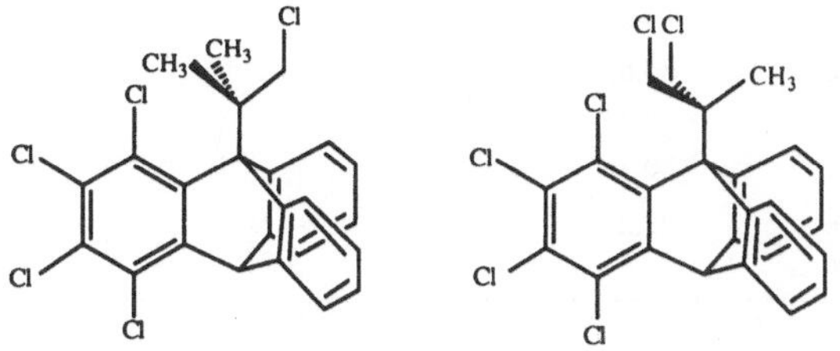

ap-**3-36**   ap-**3-37**

On the basis of the negative buttressing effect, the high barrier to rotation in 9,10-bis(1-cyano-1-methylethyl)triptycene (**3-38**) (37.7 kcal/mol) [42] can now be understood. This was a puzzling example of high barriers to rotation, because the compound does not carry a substituent at the 1-position and the cyano group is relatively small, yet the barrier was as high as that in the compounds which carry bulky substituents. The relaxation due to the absence of a 1-substituent and the small size of the cyano group must be responsible for the observed barrier to rotation. Then a question arises: what is the effect of substituting one of the methyl groups in the 9-*tert*-butyl by a cyano group on the barrier to rotation?

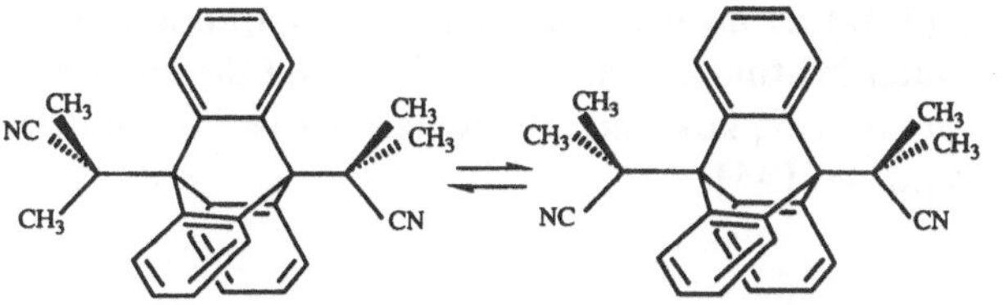

dl-**3-38**   meso-**3-38**

This can be seen in examples of compounds which carry a cyano (**3-39**) or a methoxycarbonyl group (**3-40**) in the 9-substituent. The results are shown in Table 3.13 [43]. As shown in the table, the cyano group gives a rather high barrier to rotation and the effect is comparable with that of a methyl group. Probably the stabilization of the ground state cancels the lowering of the transition state energy, although the population difference precludes a detailed discussion on the effects of the two substituents. Clearly, the methoxycarbonyl group lowers the barrier. The result is attributed to the increase in the ground state energy due to the fact that the oxygen atom of the group must penetrate into the triptycene skeleton to result in destabilization of the ground state. The enhancement of the barrier to rotation for the process, *sc* → *ap*, in compounds carrying a chloro or a methoxy substituent at the 1-position, is attributed to the

Table 3.13. Activation parameters for rotation in compounds **3-39** and **3-40**

Compound	X	Y	Process	$\Delta H^{\neq}$ /kcal mol^{-1}	$\Delta S^{\neq}$ /e.u.
3-39	Cl	Cl	$sc \rightarrow ap$	35.3	− 2.3
			$ap \rightarrow sc$	32.9	− 3.3
3-39	CH$_3$	H	$sc \rightarrow ap$	35.8	− 1.6
			$ap \rightarrow sc$	34.6	1.0
3-40	Cl	Cl	$sc \rightarrow ap$	33.5	− 3.6
			$ap \rightarrow sc$	30.3	− 4.4
3-40	CH$_3$	H	$sc \rightarrow ap$	28.2	− 14.5
			$ap \rightarrow sc$	23.4	− 20.3

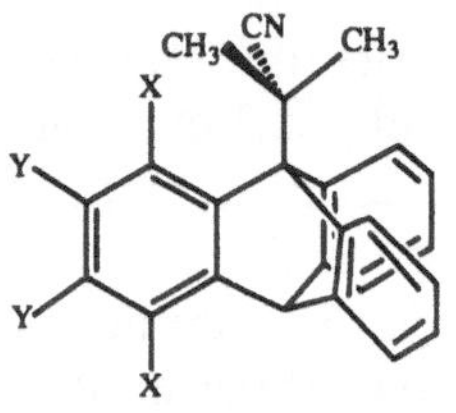

sc-**3-39** 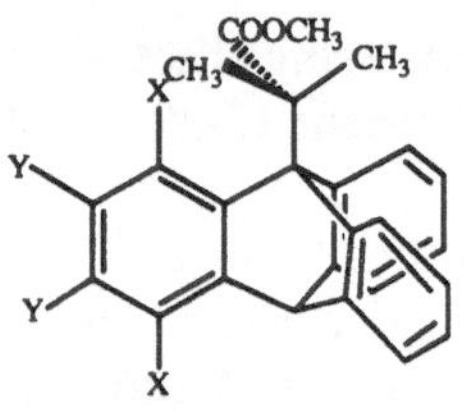   sc-**3-40**

stabilization of the ground state due to charge-transfer interactions involving the 1-substituent and the cyano or the methoxycarbonyl group.

This stabilization effect of the ground state by a small group is clearly seen by using ditriptycyl derivatives (**3-41**) as an example. These compounds are reported not to isomerize even after heating at 300 °C for 170 h and the barrier to rotation is estimated to be more than 55 kcal/mol, the highest barrier to rotation about a single bond so far reported [44].

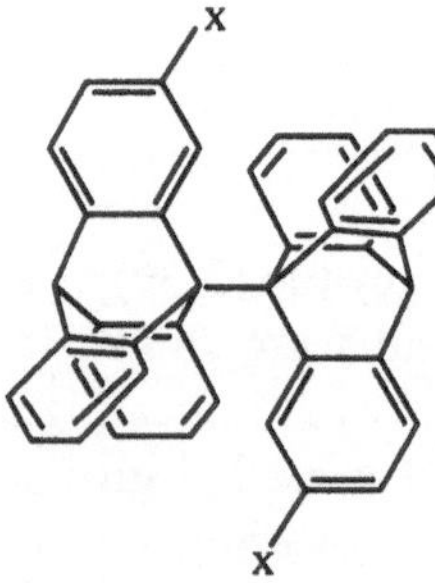

3-41

In principle, any heteroatom may be introduced in place of one of the methyl groups in the 9-*tert*-butyltriptycene derivatives for isolation of rotational isomers. However, as described in chapter 2, if one of the substituents is small in the 9-substituent in the triptycene series, the molecule tends to reside at the most sterically relaxed position. Thus compounds **3-39** and **3-40** tend to take *sc* conformations if they carry a substituent at the 1-position.

In order to overcome this problem, one could use compounds that do not carry substituents in any of the *peri*-positions. One such example is 2,3-dichloro-9-(1-hydroxy-1-methylethyl)triptycene (**3-42**). The *ap* and *sc* rotamers of this compound were isolated and the barrier to rotation was found to be 34.0 kcal/mol for the *sc* → *ap* process at 153 °C [45]. It should be noted that the barrier to rotation in compound **3-42** is reduced to some extent with respect to that of compound **3-34** which has the same substitution pattern.

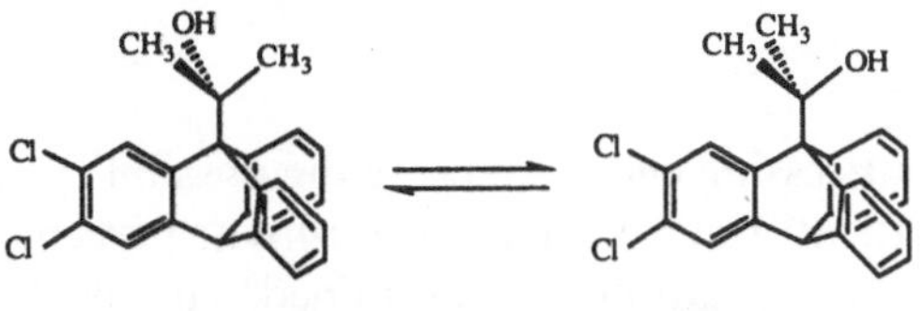

sc-**3-42**   ap-**3-42**

The triptycene skeleton is ideal for realizing a very high barrier indeed, due to the fact that the ground state energy is relatively low because of its structure: it provides notches that are made by relatively thin benzene rings and can accommodate substituents at the other end of the C_9-to-substituent bond. And yet, the transition state energy is very high because the 9-substituent must pass over the *peri*-carbons when it rotates. Thus the observed barrier to rotation is considerably enhanced in 9-substituted triptycenes with respect to other ethane derivatives.

3.3.2 Triptycenes Carrying a 9-*sec*-Alkyl Substituent

The barriers to rotation in the simplest compounds of the *sec*-alkyl series, substituted 9-isopropyltriptycenes (**3-43**), have been examined in detail by dynamic NMR methods. In this case, the *ap*-isomer is so unstable due to the steric effect relative to the *sc*-isomer that the latters only are observed, except for the 1-fluoro compound. The barriers to racemization of the *sc*-isomers are shown in Table 3.14 [46]. The barrier height gradually increases, as the size of the 1-substituent becomes large. However, the barrier height seems to saturate when the bulkiness of the substituent is increased.

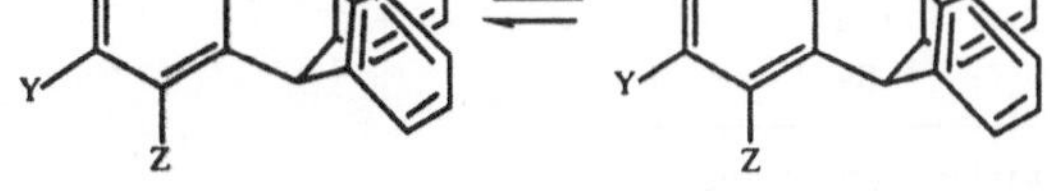

+*sc*-**3-43**   -*sc*-**3-43**

Table 3.14. Activation parameters for rotation in substituted 9-isopropyltriptycenes (**3-43**) in halogenated hydrocarbons

W	X	Y	Z	$\Delta H^{\neq}$/kcal mol^{-1}	$\Delta S^{\neq}$/e.u.	$\Delta G^{\neq}$/kcal mol^{-1}
H	CH$_3$	H	CH$_3$	17.2 ± 1.4	9.1 ± 3.7	19.9 ± 2.5
CH$_3$O	H	H	CH$_3$O	24.8 ± 1.1	3.9 ± 2.7	23.6 ± 1.9
Cl	Cl	Cl	Cl	30.1 ± 1.4	15.5 ± 2.3	25.5 ± 2.3
Br	Br	Br	Br			$23.5^{a)}$
CH$_3$	H	H	CH$_3$	21.4 ± 0.7	1.5 ± 1.7	21.8 ± 1.1

[a] $\Delta G^{\neq}$ at 175 °C. Others are at 25 °C.

This may again be the negative buttressing effect. To erase the possibility of the buttressing effects of the 2-substituent, 1-substituted 9-isopropyltriptycenes (**3-44**) were prepared and their barriers to rotation were examined. The results are shown in Table 3.15 [47].

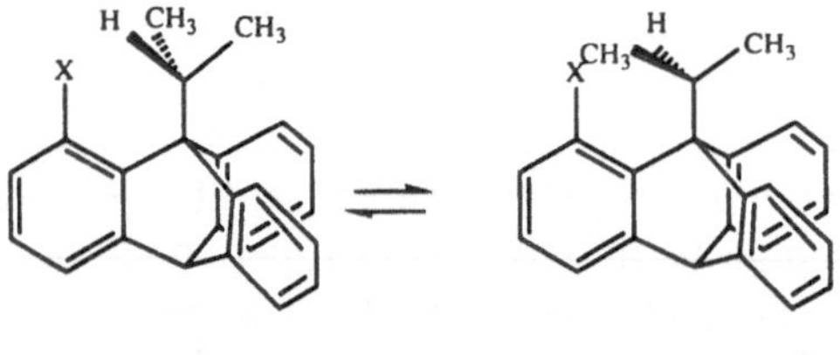

+sc-**3-44**   -sc-**3-44**

Clearly, there is again a maximum in the barrier height, if we plot it against the 1-substituent size. This time, however, the maximum is shifted to a larger substituent than in the case of *tert*-alkyl compounds: the maximum is seen at the bromo or chloro group. The slope seen in the plot is not so steep in the isopropyl compounds as in the *tert*-alkyl compounds. This must be due to the lower congestion in the former in the ground state than that in the latter because of the presence of a hydrogen in the isopropyl group instead of a methyl in a *tert*-butyl group.

The barriers to rotation in the 9-isopropyltriptycenes indicate that, if one chooses proper substituents, one should be able to manifest slow rotation about a single bond, *sec*-alkyl-to-triptycene, at room temperature. However, if one

Table 3.15. Barriers to rotation in 1-substituted 9-isopropyl-triptycenes (**3-44**)

Substituent	$\Delta G_c^{\neq}$/kcal mol^{-1}	T_c/°C
F	ca. 22	ca. 130
Cl	22.9	158
Br	22.6	153
(CH$_3$)$_3$C$^{a)}$	20.8	89

[a] This compound carries another *tert*-butyl group at the 3-position.

wishes to see the phenomenon in the isopropyl compound, one has to separate enantiomers because these compounds exist as a racemic mixture due to the steric effect. Thus compounds which form diastereomers due to restricted rotation may be designed for isolation of rotamers. They are 9-(2-acetoxy- or 2-methoxy-1-methylethyl)triptycenes **(3-45)** and 9-(1-methyl-2-propenyl)-triptycene **(3-46)**. Their dynamic processes are shown in the following schema.

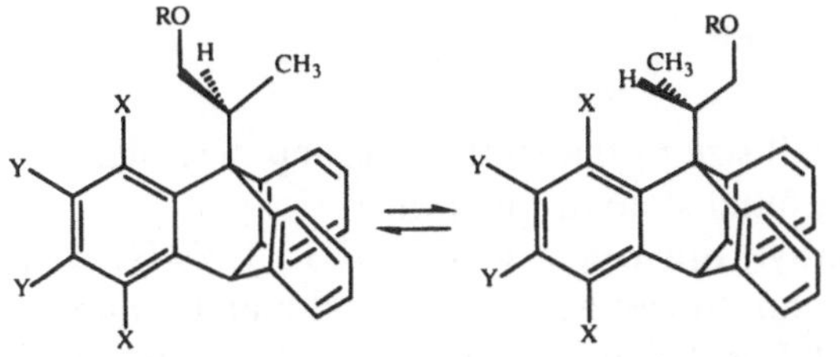

sc*(S*)-**3-45**   ap-**3-45**

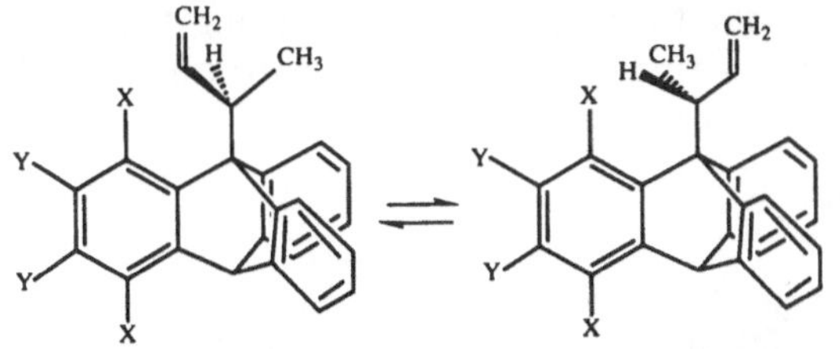

sc*(R*)-**3-46**   ap-**3-46**

The methoxy compounds (**3-45**: R=CH$_3$) which carry various substituents at 1,2,3,4-positions were isolated as *ap* isomers and they isomerized slowly at room temperature except for the 1,4-dimethoxy compound, which showed a barrier to rotation of 22.9 kcal/mol at 25 °C according to the dynamic NMR data. The barriers to rotation were found to be ca. 23.5 kcal/mol for 1,2,3,4-tetrachloro and tetrabromo compounds, which are in excellent agreement with those obtained by the dynamic NMR method [48]. As expected, the change from a methoxy to an acetoxy substituent (**3-45**: R=CH$_3$CO) did not affect the barrier height.

The 2-propenyl compounds (**3-46**) show barriers to rotation a little higher than the methoxy compound when they carry 1,2,3,4-tetrahalo substituents, ca. 24.2 kcal/mol. However, if the substituents are 1,4-dimethoxy or 1,4-dimethyl, the barriers are lower, ca. 22.5 kcal/mol. The results are attributed to both the size effect in buttressing and the relaxation of the ground state due to a smaller size of the π-system than that of a methyl [49].

The effect of substituting of a methoxy group for one of the methyls of the isopropyl group was studied. 9-(1-Methoxyethyl)triptycenes (**3-37**) showed surprisingly high barriers to rotation [51]. They are ca. 23.1 kcal/mol for 1,2,3,4-tetrahalo compounds, where the halogen is either chlorine or bromine, and 21.6 kcal/mol for the 1,4-dimethyl compound for the *sc* → *ap* process. The rather high barrier of the 1-methoxyethyl compounds with respect to isopropyl

sc*(S*)-3-47   ap-3-47

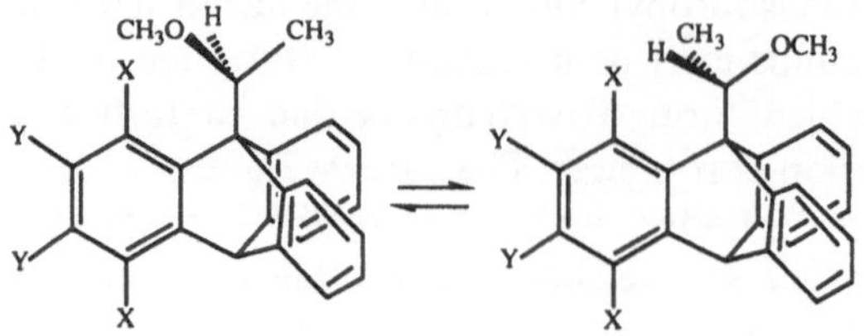

compounds irrespective to the small size of the methoxy group may again be attributed to the relaxing in the ground state in the former.

1,4-Dimethoxy-9-(1-methoxyethyl)triptycene (**3-47**: X=CH_3O, Y=H) exists as *ap* and *sc* rotamers as well and shows about 21.5 kcal/mol barriers to rotation, which are not much different from those in the 1,2,3,4-tetrahalo compounds, but 1-hydroxy-4-methoxy-9-(1-methoxyethyl)triptycene **(3-48)** showed an interesting phenomenon due to the presence of a strong hydrogen bond between the 1-hydroxy and the methoxy groups in the substituent [51].

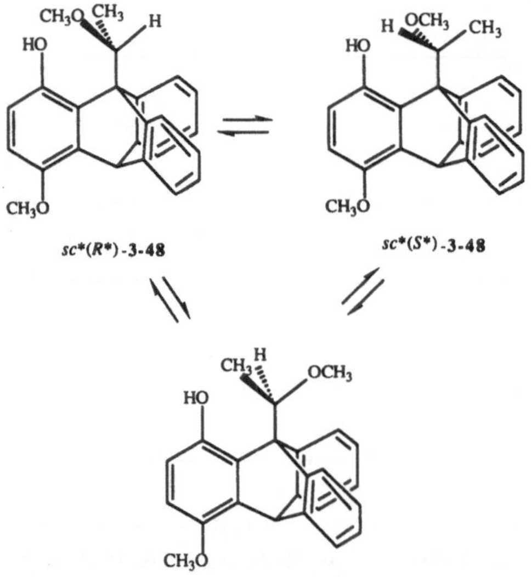

sc*(R*)-**3-48**   sc*(S*)-**3-48**

ap-**3-48**

It exists in two *sc* forms [*sc**(*R**) and *sc**(*S**)] because of the hydrogen bond in chloroform, with the former predominating. In dimethyl sulfoxide, it can form intermolecular hydrogen bonding and thus all three rotamers are found, where the *ap* rotamer predominates. The barriers to rotation are 21.0 and 21.9 kcal/mol at 112 °C for the *sc**(*R**)→*sc**(*S**) and *sc**(*S**)→*sc**(*R**) processes, respectively, in chlorobenzene. It is interesting that a barrier to rotation, in which a large substituent passes over the 1-hydroxy group in the transition state, is observed in this case.

3.3.3 Triptycenes Carrying a 9-*prim*-Alkyl Substituent

The first triptycene compound, of which dynamic behavior of the 9-substituent was studied, was 9-(chloromethyl)triptycene (**3-49**) [52]. The *peri*-proton signals show coalescence/decoalescence phenomena and the barrier to rotation of the chloromethyl group was calculated to be 13 kcal/mol. Later the barrier to rotation in 9,10-bis(chloromethyl)triptycene (**3-50**) was studied by ^{13}C NMR to find data in agreement with the mono-chloromethyl compound [53].

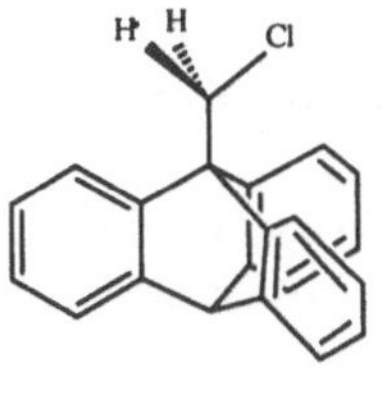

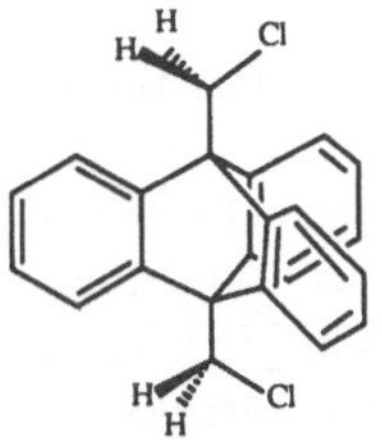

Differing from the β-substituent in the *tert*-butyl and isopropyl-triptycene derivatives, the substituent which is carried by a methyl group at the 9-position of triptycene should affect the barrier height for rotation because the substituent interacts with the peri-hydrogen at the transition state for rotation. Thus it is necessary to obtain barriers to rotation of a methyl group in order to discuss the effect of the *peri*-substituent on the barrier height for rotation in 9-*prim*-alkyltriptycenes.

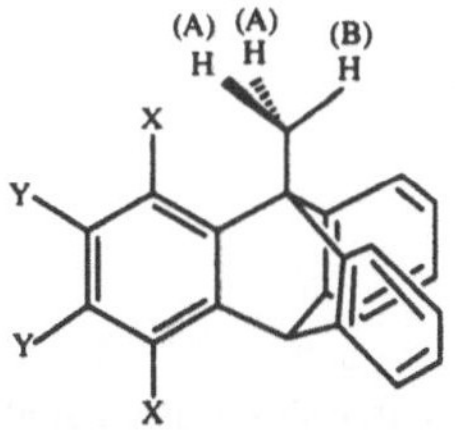

The results for 9-methyltriptycenes (**3-51** and **3-52**) are summarized in Table 3.16 [54]. It is necessary to carry substituents at two *peri*-positions for tripty-cenes in order to demonstrate the clearly resolved signals of protons in a methyl group in 9-methyltriptycenes in dynamic NMR spectroscopy at 100 MHz. However, it is also possible to obtain rough barrier heights for rotation of a methyl group, if triptycenes carry only one substituent in a peri-position, because the methyl group exhibits the A_2B type signals in this case. In addition to the substitution effects of *peri*-position(s), barriers are affected by the size of the *peri*-substituent: the larger the substituent, the higher the barrier to rotation, differing from other *tert*-alkyl and *sec*-alkyl cases. This seemed to be ascribable

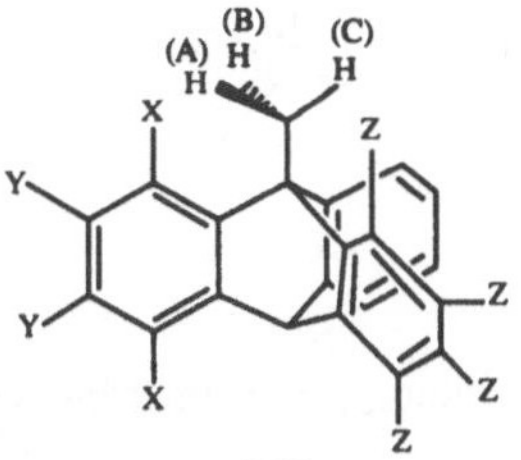

Table 3.16. Activation parameters for internal rotation in substituted 9-methyltriptycenes (**3-51** and **3-52**)

X	Y	Z	$\Delta H^{\neq}$/kcal mol^{-1}	$\Delta S^{\neq}$/e.u.	$\Delta G^{\neq}_{298}$/kcal mol^{-1}
H	H	Cl	8.6 ± 0.4	-5.9 ± 2.1	10.3 ± 1.0
H	H	Br	9.6 ± 0.4	-2.1 ± 2.0	10.2 ± 1.0
Cl	Cl	Cl	13.5 ± 0.6	9.8 ± 3.0	10.6 ± 1.0
Br	Br	Cl	12.4 ± 0.3	3.4 ± 1.3	11.4 ± 0.6
CH$_3$	H	Cl	12.1 ± 0.3	3.0 ± 1.4	11.2 ± 0.7

to the small size of hydrogen which makes the ground state congestion less important.

Recently, however, 9-methyltriptycenes (**3-53** and **3-54**) which carry substituent in all *peri*-positions were synthesized and their barriers to rotation examined. The results are shown in Table 3.17 [55]. The results indicate that, if a *tert*-butyl group is introduced into one of the *peri*-positions, the barrier to rotation is diminished, although other substituents show normal tendency in that the bulkier substituent enhances the barrier. Therefore, the negative buttressing effect is observed even in the methyl rotation if the size of the substituent surpasses a certain limit. A comparison of the results in Tables 3.16 and 3.17 also indicates that in normal cases the third *peri*-substituent enhances the barrier to rotation of the methyl group.

3-53   3-54

Table 3.17. Activation parameters for rotation in 9-methyltriptycenes that carry substituents at all *peri*-positions (**3-53** and **3-54**)

W	X	Y	Z	$\Delta H^{\neq}$/kcal mol^{-1}	$\Delta S^{\neq}$/e.u.	$\Delta G^{\neq}_{298}$/kcal mol^{-1}
				Compound **3-53**		
H	H	H	H	10.9 ± 0.2	0.2 ± 1.0	10.9
F	F	F	F	13.4 ± 0.2	3.4 ± 1.1	12.5
CH$_3$O	H	H	CH$_3$O	13.6 ± 0.6	4.2 ± 2.3	12.6
CH$_3$	H	H	CH$_3$	14.1 ± 1.3	4.1 ± 5.0	13.1
(CH$_3$)$_3$	CH	(CH$_3$)$_3$C	H	9.5 ± 2.5	2.5 ± 11.9	8.9
H	(CH$_3$)$_3$C	H	(CH$_3$)$_3$C	11.8 ± 0.2	1.2 ± 0.8	11.5
				Compound **3-54**		
CH$_3$O	H	H	CH$_3$O	12.1 ± 0.4	3.6 ± 2.0	11.2
CH$_3$	H	H	CH$_3$	13.1 ± 0.5	4.0 ± 2.2	12.1

Independently, restricted rotation of a methyl group in a triptycenequinone (**3-55**) has been reported. The barrier to rotation of the methyl group is low 7.2 kcal mol^{-1} [56]. This is an expected result because the quinone moiety is smaller than the normal 1-substituent in triptycenes and the size is reflected in the barrier height.

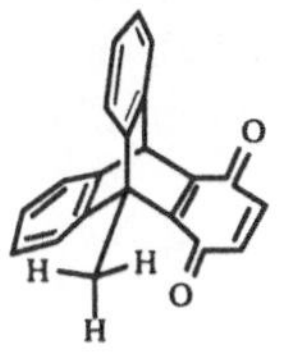

3-55

Apparently, substitution of a chlorine atom for a hydrogen in 9-methyltriptycene enhanced the barrier to rotation considerably: this is clear when one compares the results of the chloromethyl compounds (**3-49** and **3-50**) with those of the methyl compounds (**3-51** through **3-54**). It is therefore interesting to see the effect of substituent at the methyl group of methyltriptycenes on the barrier to rotation.

The barriers to rotation in 9-benzyltriptycenes (**3-56**) have been studied [40]. The barriers are ca. 12 kcal/mol for the 1,4-dimethoxy compound and ca, 14 kcal/mol for the 1,2,3,4-tetrachloro compound. The reason for rather low barriers to rotation in these compounds with respect to the 9-chloromethyl compound (**3-49** and **3-50**) must be attributed to the high ground state energy, because of the steric requirement of the phenyl group: due to the size of the benzene ring, a part of it must be placed into the notch of the triptycene skeleton to make the interaction between it and the triptycene moiety strong. However, if the phenyl group in **3-56** has to pass over the *peri*-substituent, this is another story (see p. 87, 90). Therefore, the results described here can be taken as the barriers to a path in which the benzyl benzene ring passes over the unsubstituted benzeno bridges.

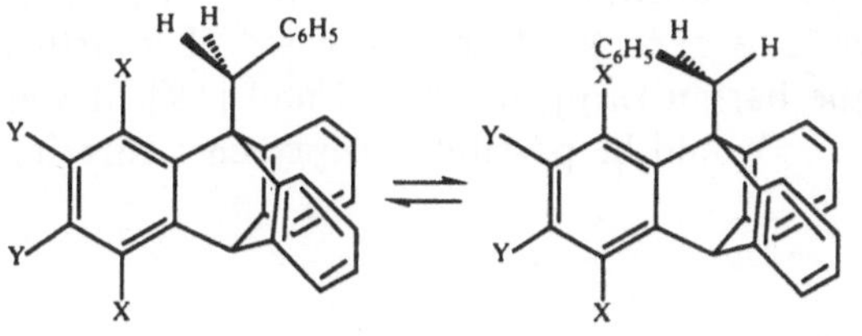

ap-**3-56**   sc-**3-56**

The barrier to rotation of an ethyl group in the 9-position was studied with use of 9-ethyl-1,4-dimethoxytriptycene (**3-57**) [57]. This compound shows the barrier of 17 kcal/mol, which is well above that of the chloromethyl compound. The enhancing effect of the methyl group on the rotational barrier about the C_9-to-substituent bond seems very large.

9-Allyl-1,4-dimethoxytriptycene (**3-58**) exhibits the barrier to rotation of ca. 13.8 kcal/mol [58]. Comparison of these values together with those in 9-benzyltriptycenes (**3-56**) indicates that the allyl group shows rather a high barrier to rotation. The barrier to the allyl rotation is certainly lower than that in the ethyl compound (**3-57**) but is higher than that in the benzyl compound (**3-56**: X=CH$_3$O, Y=H). The fact can be again attributed to the stability of the ground state of compound **3-58** relative to the benzyl compound (**3-56**). Being smaller in size, the vinyl group in compound **3-58** seems to give less interaction with the benzeno bridges in the ground state for rotation.

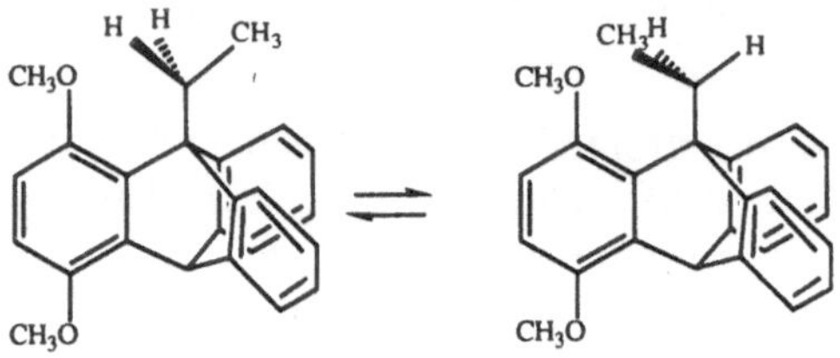

ap-**3-57**   sc-**3-57**

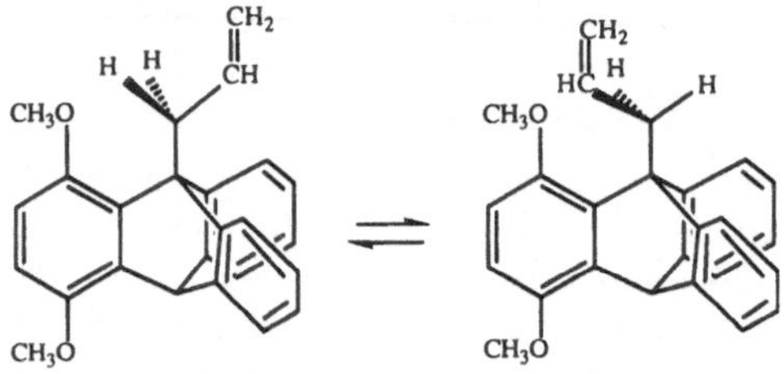

ap-**3-58**   sc-**3-58**

The barriers to rotation of *prim*-alkyl groups attached to the 9-position of triptycene seem to be too low for isolation of rotational isomers at room temperature, as far as they carry a substituent at one *peri*-position. Examination of the barrier to rotation of the 9-substituent in 9-benzyl-1,2,3,4,5,6,7,8-octachlorotriptycene (**3-59**) showed, as expected from the cases of 9-methyl derivatives, an enhancement of the barrier of up to 18 kcal/mol [59]. It was hoped that isolation of atropisomers would be possible if triptycenes carrying

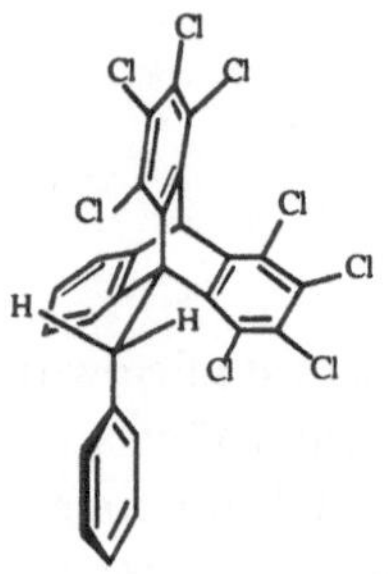

sc-**3-59** 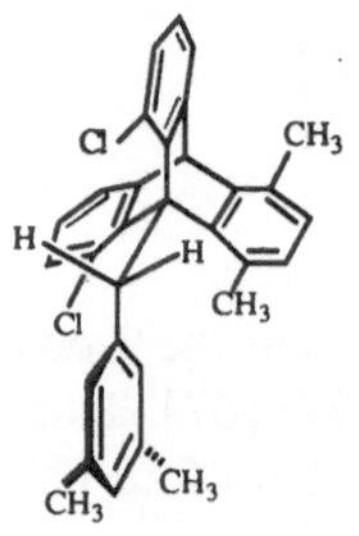   sc-**3-60** substituents at three *peri*-positions were synthesized, even if the substituent were of *prim*-alkyl type. Thus some 9-*prim*-alkyltriptycenes that carry substituents at 1,8,13 positions were synthesized. The first example was 8,13-dichloro-1,4-dimethyl-9-(3,5-dimethylbenzyl)triptycene (**3-60**) which was separated into stable rotational isomers. The barrier to rotation was 24.7 kcal mol^{-1} at 69 °C [59].

In order to see the effect of the two methyl groups in the benzyl group of compound **3-60** on the barrier to rotation, 9-benzyl-8,13-dichloro-1,4-dimethyltriptycene (**3-61**) was synthesized. The barriers to rotation were ca. 25.3 kcal/mol for both *ap* → *sc* and *sc* → *ap* processes at 54 °C [58]. Thus this compound was also separated into rotational isomers at room temperature. The methyl group at the 1-position is very important in enhancing the barrier because substitution of a methoxy group for the 1,4-dimethyl in compound **3-60** lowers the barrier to 20 kcal/mol to make it impossible to isolate the atropisomers at room temperature [59]. The removal of the 1,4-dimethyl groups gives the barrier to rotation of 16.1 kcal/mol which is a little lower than that in the octachloro compound (**3-59**) probably because of the buttressing effect.

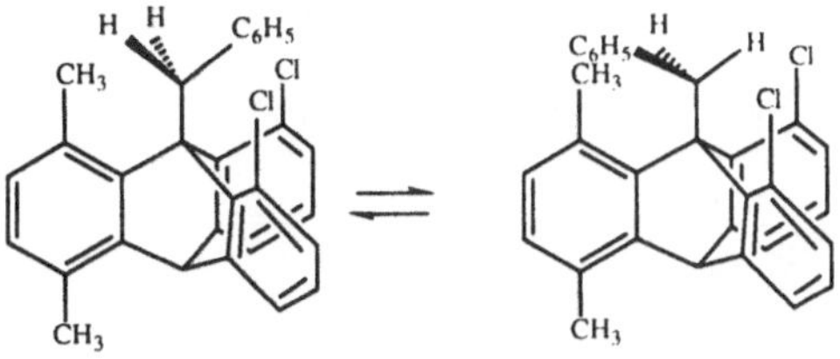

ap-**3-61**   *sc*-**3-61**

Here, the group attached to the methyl in the 9-position is important in determining the barrier height, because when the benzyl group in compound **3-61** is replaced by either an ethyl or an allyl, the barrier in these compounds (**3-62**) becomes 20 or 22 kcal/mol, respectively. The high barriers in the π-system may be attributed to the release of strain in the ground state. The high barrier for the benzyl compound is interesting. It is ascribed to the high energy of the transition state for rotation, when the benzylic phenyl group move past one of the *peri*-substituents.

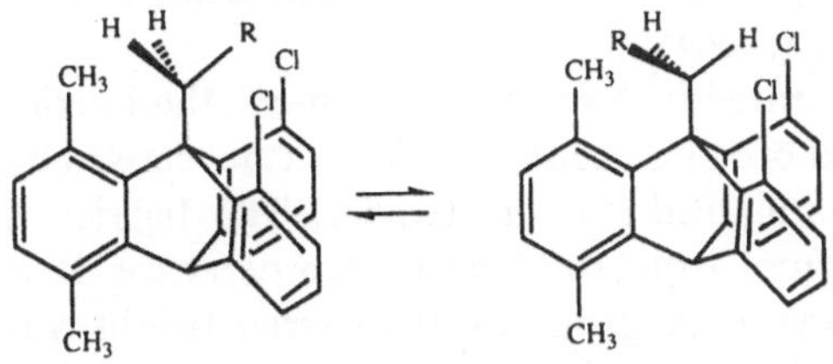

ap-**3-62**   *sc*-**3-62**

3.3.4 Atropisomers About a Carbon-to-Heteroatom Bond

Atropisomers of 9-substituted triptycenes that carry a substituent connected to the 9-position with a heteroatom are now known. Barriers to rotation in these compounds are governed by the bond lengths and the ease of angle deformation. The angle deformation is known to become easier as the heteroatom becomes large, if one compares them in a given column of the periodic table [60].

As a natural extension of the carbon-to-carbon compound, it will be interesting to compare the barrier to rotation in carbon-to-silicon compounds. Examination of the NMR spectra of 1-substituted 9-(trimethylsilyl)triptycene at high temperatures revealed that the barrier to rotation about the C_9–Si bond is in excess of 25 kcal/mol. In order to make the atropisomers possible, a benzyl group was introduced in place of one of the methyls in the trimethylsilyl group. The rotational isomers were partially purified and it was found that the compound (3-63) slowly isomerized at 120°C. Although the barrier is not calculated, it should be about 31 kcal/mol [61]. The barrier height is considerably reduced from the value in the corresponding C–C compound.

ap-3-63   *sc*-3-63

The reduction is attributed to two factors. One is the longer bond length of C–Si than C–C. Another is the easier bond angle deformation of the C–Si–C. The extents of the roles of the factors are not known but the latter seems to play an important role as discussed later. Although the ground state energy of these rotamers is expected to be lower than the C–C compound (3-34), the effect of the ground state energy seems to be small.

Another possibility of introducing a substituent which has a heteroatom connected to C_9 of the triptycene system and bears three substituents on the heteroatom is a quaternary ammonium salt or its equivalents. No quaternary salts are reported, because they are easily attacked by nucleophiles due to the steric strain in the ground state but a tertiary amine oxide was successfully separated into *ap* and *sc* isomers [62]. They are isomers of *N,N*-dimethyl-*N*-(2,3-dimethyl-9-triptycyl) amine oxide (3-64).

Although this compound was separated into rotational isomers, the barrier to isomerization could not be obtained because decomposition sets in, caused by heating which was necessary for examination of the barrier. The barrier is estimated to be in excess of 30 kcal/mol from the heating temperature. It is unfortunate that the effect of the amine oxide group on the barrier height was not measurable. This point will be discussed in Sect. 3.4.

Attempted synthesis of 1,4-dimethoxy-9-(dimethylamino)-triptycene *N*-oxide was unsuccessful, apparently due to the steric effect. Thus compound **3-64** is suitable for studying the isolation of rotational isomers in this class of compounds owing to the lack of the 1-substituent which raises the ground state energy more severely than the carbon-to-carbon bond case (**3-34**) thanks to the N–C bond being shorter than C–C.

ap-**3-64**   sc-**3-64**

Rotational barriers about a C–O bond are very low. 9-Methoxytriptycene derivatives must carry substituents in three *peri*-positions to show restricted rotation in the NMR spectra. 8,13-Dichloro-9-methoxy-1,4-dimethyltriptycene (**3-65**) showed two signals due to the methoxy protons at − 100°C at 400 MHz and the dynamic NMR study of this compound showed the barrier to rotation to be 11.4 kcal/mol, for the *ap* → *sc* process [63]. This low barrier is attributed to two factors, short C–O bonds that raise the ground state energy and the small size of the lone pair electrons that lower the transition state energy, with respect to the corresponding 9-ethyl compound.

ap-**3-65**   sc-**3-65**

However, if a triptycene derivative carries an *o*-substituted phenoxy group at the 9-position, the barrier to rotation about the C_9-to-substituent bond is considerably enhanced. A substituent at only one *peri*-position suffices to detect the restricted rotation by NMR spectroscopy. 1,4-Dimethyl-9-(*o*-tolyloxy)triptycene (**3-66**) is stable only in conformations in which the *o*-methyl group is positioned on the outside with respect to the triptycene skeleton. Therefore, the isomerization scheme is simply written as shown on the next page.

If one considers the mechanism of isomerization, one notices that the direct isomerization as it is depicted in the scheme on the next page is not possible. If rotation of the *o*-tolyl group in the *ap* conformation takes place to form the *sc* isomer, the methyl group is positioned on the inside. This conformation is very unstable. Therefore, it isomerizes quickly to another conformation by rotation in which the methyl group is directed to the outside. Yet the rotation of the

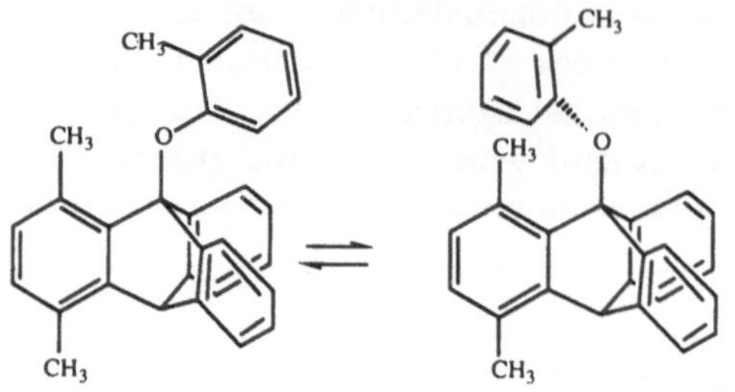

ap-3-66   sc-3-66

o-tolyl group in a given notch is not possible due to the steric effect. Therefore, the *ap → sc* conversion is possible only when the o-tolyl group passes over the l-methyl group. The reverse process is also the same, though it is possible that the + *sc* → − *sc* process takes place without passing over the methyl group. This makes the isomerization energy high. It is ca. 18 kcal/mol for the *ap → sc* process and 10.5 kcal/mol for the + *sc* → − *sc* process [64].

This is a kind of molecular gear. The present example is the case of two-toothed and three-toothed bevel gears. It is an interesting phenomenon but is outside the scope of this book. Interested readers are recommended to read the review by Iwamura and Mislow [65].

Sulfur can be made tetrahedral and thus is expected to be able to give a higher barrier to rotation than oxygen does if it is introduced to the 9-position of the triptycene skeleton. 9-Benzylsulfonyl-1,4-dimethoxytriptycene (**3-67**) was separated into rotational isomers. The barrier to isomerization is 32 kcal/mol [66]. The same literature implies that the corresponding sulfoxide is also a mixture of rotational isomers but separation of the isomers was not successful.

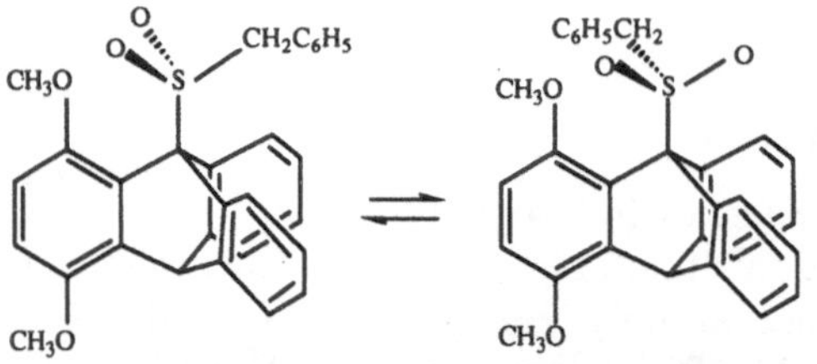

ap-**3-67**   sc-**3-67**

3.4 High Barriers to Rotation in Miscellaneous Compounds

A number of scattered reports can be found in the literature which describe high barriers to rotation. Here, a limited number of example will be described with comparison of areas of interest mentioned in the previous sections.

3.4.1 Atropisomerism About an sp^3–sp^3 Bond

As already mentioned, at the outset, isolation of atropisomers was achieved with use of a dihydroethenoanthracene derivative (3-68), which carried two methoxycarbonyl groups on the etheno bridge. The barrier to isomerization was 32 kcal/mol [67]. Thus, substitution of a bis(methoxycarbonyl)etheno bridge for an o-benzeno bridge in the triptycene skeleton reduces the barrier by about 10 kcal/mol.

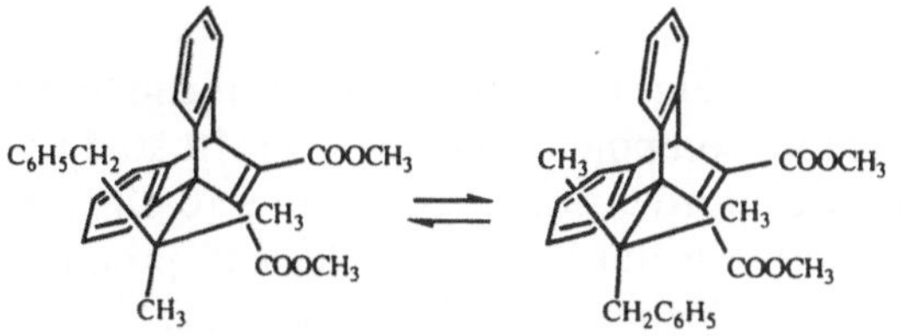

ap-3-68   sc-3-68

Then it is interesting to see how the additional benzeno bridge affects the barrier. Unfortunately, the barrier height for the compound which lacks the methoxycarbonyl group has never been measured but a dynamic NMR study on 9-tert-butyl-1,2,3,4-tetrachloro-9,10-dihydro-9,10-ethenoanthracene (3-69) indicated that it is in excess of 25 kcal/mol because the two peaks attributable to the methyl protons in the tert-butyl group remained intact even at 200°C [68]. This is compared with that of the corresponding tetrafluoro compound (3-70, 20 kcal/mol) [69]. Apparently, the benzeno bridge together with the chloro substituent enhanced the barrier height considerably.

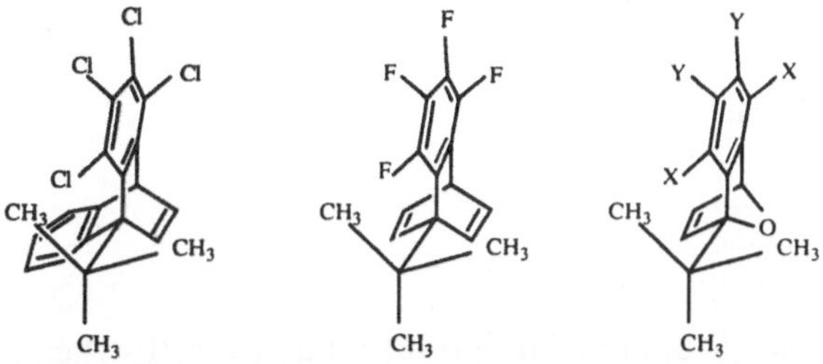

3-69   3-70   3-71

The height of the barrier to rotation in compounds (3-71) which possess an oxido bridge instead of the etheno has been studied. The results are shown in Table 3.18 [70]. The barriers are considerably lower than the cases of triptycene derivatives. This is attributed to the small size of the oxido bridge. It is interesting that one also sees a maximum barrier height when the barriers are plotted against the size of the substituent. The shift of the maximum to the chlorine in this case, instead of the fluoro compound in triptycene derivatives, undoubtedly suggest that the steric requirement of the oxido bridge is responsible for this fact.

Table 3.18. Activation parameters for rotation of *tert*-butyl group in substituted 1,4-oxido-1-*tert*-butyl-1,4-dihydronaphtalenes (**3-71**) in CS_2–CH_2Cl_2

X	Y	$\Delta H^{\neq}$/kcal mol^{-1}	$\Delta S^{\neq}$/e.u.	$\Delta G_{298}^{\neq}$/kcal mol^{-1}
H	H	7.1 ± 0.6	-9.4 ± 2.3	9.9 ± 1.0
CH$_3$O	H	10.2 ± 0.5	-4.1 ± 1.0	11.4 ± 0.8
F	F	7.6 ± 1.0	-11.8 ± 1.0	11.1 ± 1.0
Cl	Cl	10.1 ± 0.6	-6.5 ± 1.0	12.0 ± 0.9
Br	Br	12.4 ± 0.4	4.5 ± 1.6	11.1 ± 0.8

The effect of changing the *o*-benzeno bridge in triptycene to an etheno bridge has also been studied in the heteroatom compounds. Dynamic NMR study on the barrier to rotation in 1,2,3,4-tetrachloro-9-trimethylmetallo-9,10-dihydro-9,10-ethenoanthracenes (**3-72**), where the metal is the IVa element, produced results that are shown in Table 3.19 which also includes the data for the corresponding carbon compounds [68]. It can be seen that, as one descends the periodic table, the barrier to rotation is considerably lowered. This is again attributed to the bond lengths and the ease of angle deformation [60]. It is noteworthy that, the barrier to rotation is lowered, even though the molecular model indicates that a severe interaction is expected between the *peri*-substituent and the CH$_3$–M group in the transition state for rotation due to the long C–M bond length. This will mean that, in these cases, the ease of angle deformation is important.

3-72   *ap*-3-73

The barrier to rotation in 9-dimethylamino-9,10-dihydro-9,10-etheno-anthracene *N*-oxide (**3-73**) has been determined. This compound crystallized in

Table 3.19. Activation parameters for rotation of trimethylmetallo group in 9-trimethylmetallo-1,2,3,4-tetrachloro-9,10-etheno-9,10-dihydroanthracene (**3-72**) in halogenated hydrocarbons or CS$_2$

M	$\Delta H^{\neq}$/kcal mol^{-1}	$\Delta S^{\neq}$/e.u.	$\Delta G_{298}^{\neq}$/kcal mol^{-1}
C			> 25
Si	20.1 ± 0.5	0.6 ± 1.8	19.9 ± 0.5
Ge	14.7 ± 0.7	-8.4 ± 2.1	17.2 ± 1.2
Sn	12.6 ± 0.4	2.2 ± 0.5	11.7 ± 0.4

the *ap*-form and on dissolution undergoes slow isomerization to the *sc*-form. The barrier to isomerization is 24.3 kcal/mol [71]. The barrier height is low relative to the triptycene derivative as expected.

However, it is noted that the barrier in this compound is much higher than the corresponding 9-*tert*-butyl-1,2,3,4-tetrafluoro-9,10-dihydro-9,10-etheno-naphthalene (**3-70**). This indicates that the oxidodimethylamino group in this system rotates more sluggishly than a *tert*-butyl group. It is ascribable to the short length of the N–C bond in addition to a factor that **3-73** possesses a benzeno bridge in place of an etheno bridge in **3-70**. These factors in **3-73** seem to be more effective in enhancing the barrier to rotation.

3.4.2 Atropisomerism About an sp^3–sp^2 Bond

As a natural extension of the study of the barrier height about a triptycene-to-substituent bond, barriers to rotation in triptycene compounds which carry an aromatic substituent at the 9-position were examined. Due to the congestion in the ground state, restricted rotation about the C$_9$-to-aromatic bond was not observed, if the aryl group is an unsubstituted phenyl: the barrier is very low. If the aromatic compound is substituted at only one of the *o*-positions, the change in the NMR spectra could be observed at ambient or at temperatures a little lower than that. Thus the barrier to rotation is ca. 13–15 kcal/mol both for *o*-methoxy and *o*-methyl compounds (**3-74**) [72]. Here again, if the 1-substituent is a methyl, the barrier to rotation is lower by ca. 1 kcal/mol than the corresponding methoxy compound.

Here, there is a question about the stable conformation of these compounds. The following pieces of evidence led the investigators to postulate that the stable conformations are those in which the *o*-substituent and the *peri*-substituent are close to each other as are shown in the following scheme. Firstly, if the 1-substituent is a fluorine, a long range coupling between the fluorine and the *o*-substituent of the 9-aryl group is observed. Secondly the barrier to rotation is relatively unaffected by the 1-substituent. And thirdly, there is no NMR signal due to aromatic protons which are located at a high field. If the ground state conformation were *ap*, a signal due to a *peri*-proton at a high field should be expected. Thus it is concluded that *ap*-conformations are so unstable that they are not observed experimentally.

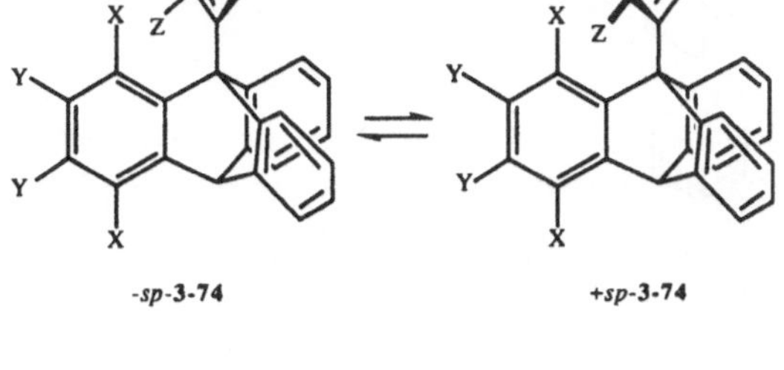

-sp-**3-74**   +sp-**3-74**

The proposed dynamic process observed here are shown in the following scheme with use of Newman-type projections. The ground state is the near-*sp* (*sp*-**3-75**) and the rate determining step is the passing of the *o*-substituent over one of the unsubstituted benzeno bridge. Passing of the *o*-substituent over the *peri*-substituent of triptycene does not occur. Conformations in which the *o*-substituent takes the position of *ac* and conformations which are near-*ap* (*ap*-**3-75**), are too unstable to detect and they thus intervene as unstable intermediates.

In accordance with this assignment, 1,2,3,4,5,6,7,8-octachloro-9-(*o*-methoxy- or methylphenyl)triptycene failed to show any dynamic processes in the dynamic NMR study. Here the ground state is very unstable due to the fact that avoiding the steric interaction by rotation is not possible because of the presence of two tetrachlorobenzeno bridges. Thus the internal rotation is not frozen until very low temperatures are reached.

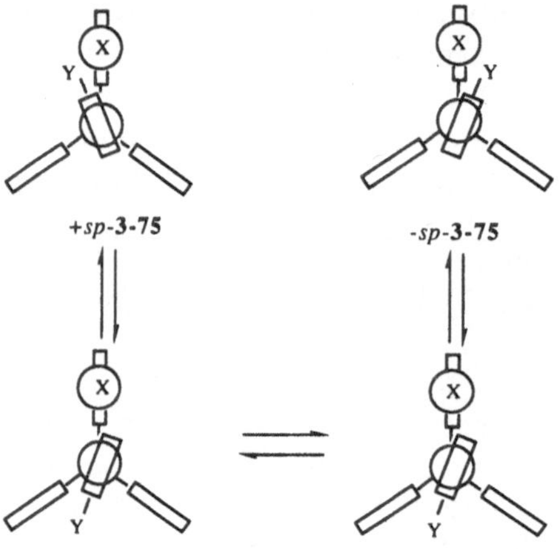

+*sp*-**3-75**   -*sp*-**3-75**

+*ap*-**3-75**   -*ap*-**3-75**

Although 9-arylfluorenes exhibit a barrier to rotation about the C_9-to-aryl bond high enough to permit isolation of rotamers at room temperature, its analog, 9-mesitylxanthene (**3-76**) shows a rather low barrier to rotation [73]. This is because the central oxin ring is not planar but a shallow boat. Because of this nature, the ring flip is very easy and the mesityl ring can take either axial (axial **3-76**) or equatorial (equatorial **3-76**) positions. Then the mesityl group is

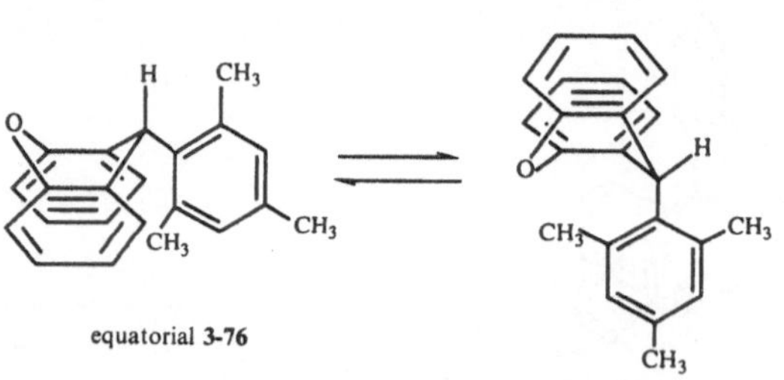

equatorial **3-76**

axial **3-76**

considered to rotate in the axial conformation, in which the steric hindrance to rotation is much smaller than that in the equatorial conformation, as is seen in the scheme on the preceding page. This conformational change is not possible in the 9-arylfluorene system because of the rigidity of the structure.

To achieve high barriers to rotation in this system, it is necessary to make the barrier to rotation in the mesityl-axial form high or to make this conformation very unstable. This is not possible in the xanthene system but is so in dihydroanthracene system (3-77) by taking advantage of interactions with substituents at the 10 position.

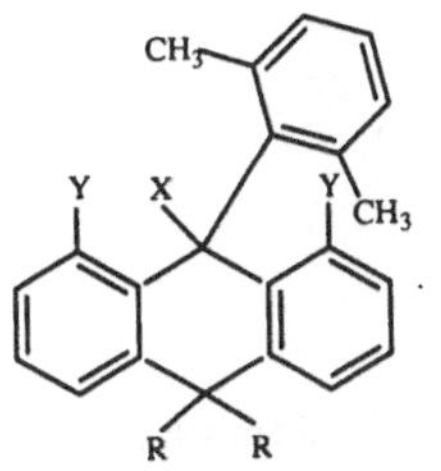

3-77

Though the introduction of two methyl groups in the 10 position of 9,10-dihydro-9-(2,6-xylyl)anthracene (3-78: R=CH$_3$) did not enhance the barrier to rotation of the xylyl group effectively, that of two benzyl groups (3-78: R=C$_6$H$_5$CH$_2$) did considerably, as is shown in Table 3.20 [74]. In this case, interestingly, the barrier to rotation of the xylyl group is higher for the 9-OH compound than in the 9-H compound, if they carry 1,8-dichloro substituents, which is in sharp contrast to the case of 9-arylfluorenes.

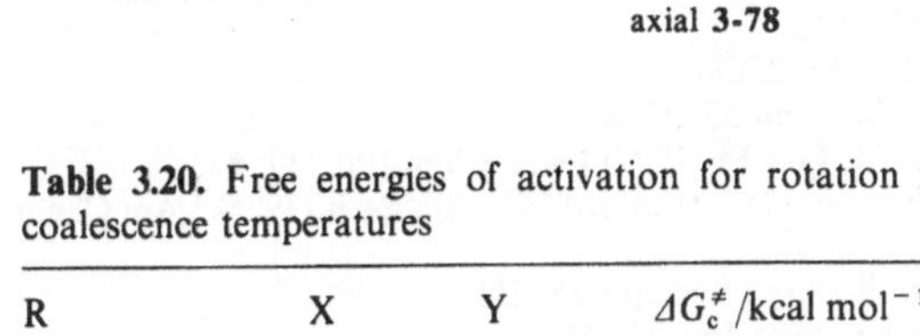

equatorial 3-78

axial 3-78

Table 3.20. Free energies of activation for rotation in compound 3-77 at coalescence temperatures

R	X	Y	$\Delta G_c^{\neq}$/kcal mol^{-1}	T_c/°C
CH$_3$	OH	H	15.4	48
CH$_3$	H	H	19.6	126
CH$_3$	OH	Cl	21.6	172
CH$_3$	H	Cl	15.0	36
C$_6$H$_5$CH$_2$	OH	H	19.6	138
C$_6$H$_5$CH$_2$	OH	Cl	> 22.7	> 200

The phenomena were interpreted as follows. If two benzyl groups are introduced to the 10 position, the phenyl groups of the benzyl groups take a conformation in which they are as far apart as possible, to make the steric effect greater for the 9-substituent. Introduction of 1,8-dichloro groups will prevent the xylyl group from rotating in the equatorial conformation. On one hand, the OH equatorial conformation would not be stable because of the parallel arrangement of three dipoles, two C–Cl and a C–O bond. On the other, if the OH group is axial, it is possible that the OH-π interaction exists between that group and the benzylic phenyl group. Taking into account these interactions, one concludes that the rotation of the mesityl group takes place in the axial conformation rather than in an equatorial one.

3.5 References

1. Ōki M (1985) Applications of dynamic NMR spectroscopy to organic chemistry, VCH, Deerfield Beach, p 43
2. Gutowsky HS, Holm CH (1956) J Chem Phys 25: 1228
3. Kurland RJ, Rubin MB, Wise WB (1964) J Chem Phys 40: 2426. Ōki M, Iwamura H, Hayakawa N (1964) Bull Chem Soc Jpn 37: 1865
4. Alexander, S (1962) J Chem Phys 37: 967. Ramey KC, Louick DJ, Whitehurst PW, Wise WB, Mukhujee R, Moriarty R (1971) Org Magn Reson 3: 201
5. Rogers MT, Woodbrey JC (1962) J Phys Chem 66: 540
6. Kleier DA, Binsch G; QCPE Program No. 165
7. Stephenson DS, Binsch G; QCPE Program No. 365
8. Forsén S, Hoffman RA (1963) Acta Chem Scand 17: 1787
9. Campbell ID, Dobson CM, Ratcliffe RG, Williams RJP (1978) J Magn Reson 29: 397
10. Dalling DK, Grant DM, Johnson LF (1971) J Am Chem Soc 93: 3678
11. Mann BE (1976) J Magn Reson 21: 17
12. Willem R (1987) Prog NMR Spectrosc 20: 1
13. Bodenhausen G, Ernst RR (1982) J Am Chem Soc 104: 1304
14. Nakamura M, Nakamura N, Ōki M (1977) Bull Chem Soc Jpn 50: 2986
15. Nakamura M, Nakamura N, Ōki M (1980) Chem Lett 605
16. Nakamura M, Kihara H, Nakamura N, Ōki M (1979) Org Magn Reson 12: 702
17. Kost D, Carlson EH, Raban M (1971) J Chem Soc, Chem Commun 656
18. Siddall TH III, Stewart WE (1969) J Org Chem 34: 233
19. Ford WT, Thompson TB, Snoble KAJ, Timco JM (1975) J Am Chem Soc 97: 95
20. Nakamura M, Ōki M (1974) Tetrahedron Lett 505. Murata S, Kanno S, Tanabe Y, Nakamura M, Ōki M (1982) Bull Chem Soc Jpn 55: 1522
21. Saito R, Ōki M (1982) Bull Chem Soc Jpn 55: 2508
22. Nakamura M, Ōki M (1980) Bull Chem Soc Jpn 53: 3248
23. Moriyama K, Nakamura N, Nakamura M, Ōki M (1987) Gazz Chim Ital 117: 655
24. Chien SL, Adams R (1934) J Am Chem Soc 56: 1787. Hanford WE, Adams R (1935) J Am Chem Soc 57: 1592
25. ˙oki M, Nakamura M, Ōki M (1982) Bull Chem Soc Jpn 55: 2512
26. Murata S, Mori T, Ōki M (1984) Bull Chem Soc Jpn 57: 1970
27. Fujisaki S, Fujimoto M, Fujii N, Umeno M, Kajigaeshi S (1979) Nippon Kagaku Kaishi 739
28. Mori T, Ōki M (1981) Bull Chem Soc Jpn 54: 1199
29. Chandross EA, Sheley CF Jr. (1968) J Am Chem Soc 90: 4345
30. Rieker A, Kessler H (1969) Tetrahedron Lett 1227
31. Albert K, Rieker A (1977) Chem Ber 110: 1804
32. Ōki M, Tsukahara J, Sonoda Y, Moriyama K, Nakamura N (1988) Bull Chem Soc Jpn 61: 4303
33. Ōki M, Otake K, Shionoiri K, Ono M, Toyota S (1991) Chem Lett 597

34. Saito R, Ōki M (1982) Bull Chem Soc Jpn 55: 3273
35. Moriyama K, Nakamura M, Nakamura N, Ōki M (1989) Bull Chem Soc Jpn 62: 485
36. Nakamura M, Ōki M (1976) Chem Lett 651
37. Nakamura M, Kihara H, Ōki M (1976) Tetrahedron Lett 1207
38. Yamamoto G, Suzuki M, Ōki M (1983) Bull Chem Soc Jpn 56: 306
39. Yamamoto G, Suzuki M, Ōki M (1983) Bull Chem Soc Jpn 56: 809
40. Suzuki F, Ōki M (1975) Bull Chem Soc Jpn 48: 596
41. Toyota S, Ōki M (unpublished work)
42. Iwamura H (1973) J Chem Soc, Chem Commun 232
43. Otsuka S, Mitsuhashi T, Ōki M (1979) Bull Chem Soc Jpn 52: 3663
44. Schwartz LH, Koukotas C, Kukkola P, Yu CS (1986) J Org Chem 51: 997
45. Ōki M, Tanaka Y, Yamamoto G, Nakamura N (1983) Bull Chem Soc Jpn 56: 302
46. Suzuki F, Nakanishi H Ōki M (1974) Bull Chem Soc Jpn 47: 3114. See also Ref 43.
47. Yamamoto G, Ōki M (1983) Bull Chem Soc Jpn 56: 2082
48. Suzuki M, Yamamoto G, Kikuchi H, Ōki M (1981) Bull Chem Soc Jpn 54: 2383
49. Kikuchi H, Hatakeyama S, Yamamoto G, Ōki M (1981) Bull Chem Soc Jpn 54: 3832
50. Tanaka Y, Yamamoto G, Ōki M (1983) Bull Chem Soc Jpn 56: 3023
51. Yamamoto G, Tanaka Y, Ōki M (1983) Bull Chem Soc Jpn 56: 3028
52. Sergeyev NM, Abdulla KF, Skvarchenko VR (1972) J Chem Soc, Chem Commun 368
53. Grishin YK, Sergeyev NM, Subbotin OA, Ustynyuk YA (1973) Mol Phys 25: 297
54. Nakamura M, Ōki M, Nakanishi H, Yamamoto O (1974) Bull Chem Soc Jpn 47: 2415
55. Yamamoto G, Ōki M (1990) Bull Chem Soc Jpn 63: 3550
56. Anderson JE, Rawson DI (1973) J Chem Soc, Chem Commun 830
57. Nakanishi H, Yamamoto O (1978) Bull Chem Soc Jpn 51: 1777
58. Yamamoto G, Ōki M (1978) Bull Chem Soc Jpn 57: 2219
59. Yamamoto G, Ōki M (1978) Angew Chem Int Ed Engl 17: 518
60. Ouellette RJ (1972) J Am Chem Soc 94: 7674
61. Nakamura N, Kohno M, Ōki M (1982) Chem Lett 1809 and private communication of Nakamura N
62. Nakamura N (1982) Chem Lett 1614
63. Yamamoto G, Ōki M (1985) Bull Chem Soc Jpn 58: 1690
64. Yamamoto G, Ōki M (1985) Bull Chem Soc Jpn 58: 1953
65. Iwamura H, Mislow K (1988) Acc Chem Res 21: 175
66. Nakamura N, Ōki M (1984) Chem Lett 143
67. Yamamoto G, Ōki M (1972) Chem Lett 45
68. Suzuki F, Ōki M, Nakanishi H (1973) Bull Chem Soc Jpn 46: 2858
69. Brewer JPN, Heaney H, Marples BA (1967) J Chem Soc, Chem Commun 27. Brewer JPN, Eckard IF, Heaney H, Marples BA (1968) J Chem Soc C 664
70. Nakamura M, Ōki M, Nakanishi H (1974) Tetrahedron 30: 543
71. Kakizaki F, Nakamura N, Ōki M (1983) Chem Lett 81
72. Nakamura M, Ōki M (1978) Bull Chem Soc Jpn 48: 2106
73. McKinley SV, Grieco PA, Young AE, Freedman HH (1970) J Am Chem Soc 92: 5900
74. Nakamura M, Ōki M (1980) Bull Chem Soc Jpn 53: 2977

4 Reactivity of Rotational Isomers

In this chapter, various findings on the reactivity of rotational isomers, including the rates of reactions, products, and reaction mechanisms, will be reviewed. The readers will find examples that convince them to think that the rotational isomers are really different compounds if they are diastereomers. The examples will further illustrate that a pair of rotational isomers are excellent for showing a detailed facet of reaction mechanisms and weak molecular interactions that affect the fate of unstable reaction intermediates.

4.1 Reactivity of 9-Arylfluorenes

As already mentioned in the foregoing chapters, the expected number of rotational isomers for the 9-arylfluorene system is six. However, due to the instability of rotational isomers that suffer from severe steric repulsion, only a pair of rotamers are isolated. Their stereostructures with stereochemical nomenclatures can be shown as follows (**4-1**).

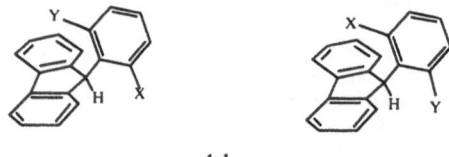

4-1

| *sp* | if X precedes Y in the sequence rule | *ap* |
| *ap* | if Y precedes X in the sequence rule | *sp* |

4.1.1 Additions and Eliminations

4.1.1.1 Additions to Olefins

A typical example of showing difference in reactivity of rotational isomers of the 9-arylfluorene series was found even in the preparation of *sp*- and *ap*-9-(2-vinyl-1-naphthyl)fluorenes (**4-3**) by acid-catalyzed elimination of water from 9-[2-(1-hydroxyethyl)-1-naphthyl]fluorene (**4-2**). Whereas the *ap*-olefin (*ap*-**4-3**) was formed smoothly by dehydration of the *sp*-alcohol, the *sp*-olefin (*sp*-**4-3**) is easily contaminated by a cyclized product (**4-4**), when sulfuric acid was used as a dehydrating reagent of the *ap*-alcohol [1].

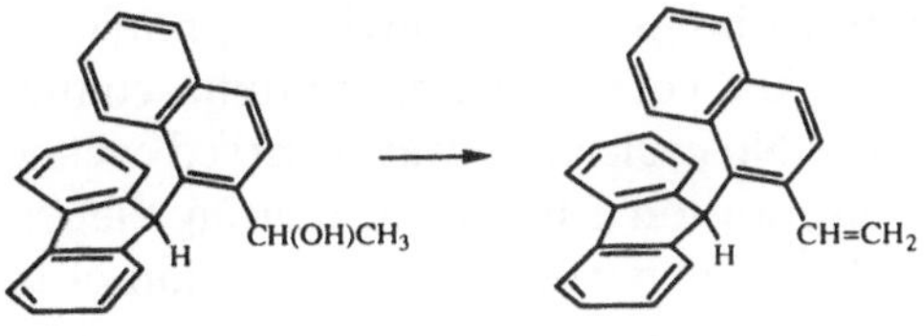

sp-4-2   ap-4-3

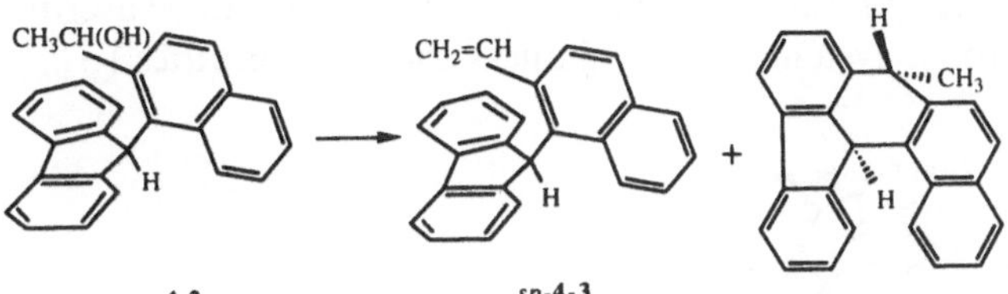

ap-4-2   sp-4-3

4-4

The results suggest that a cation (4-5) formed as a result of an S_N1 type reaction by the action of sulfuric acid on the alcohol or as a result of addition of a proton by the action of the acid on the olefin can react with the benzene ring of the fluorene moiety to give the cyclized compound. Indeed, sp-4-3 gives the cyclized product (4-4) in an excellent yield, when it is treated with concentrated sulfuric acid. On the other hand, the cation formed from the ap-olefin (4-2) is not possible to cyclize due to the structure. Thus unless it polymerizes by inter-molecular reactions, the cation survives and gives the starting material once more. This point will be further elaborated in Sect. 4.1.2.

The structure of the cyclized product (4-4) was established by both X-ray crystallography [2] and the use of the long range couplings between the 9-H and the proton attached to the carbon atom carrying the methyl group in ^1H NMR spectroscopy. The formation of this compound is rationalized by considering the intervening cation. Although this cation (4-5) can take two conformations in principle, one conformation [sp*(S*)-4-5] which possesses the methyl group over the fluorene ring will be considerably unstable due to the steric effect of the methyl group with respect to the conformation [sp*(R*)-4-5] in which the hydrogen is over the fluorene ring, immediately before the cyclization to the benzene ring.

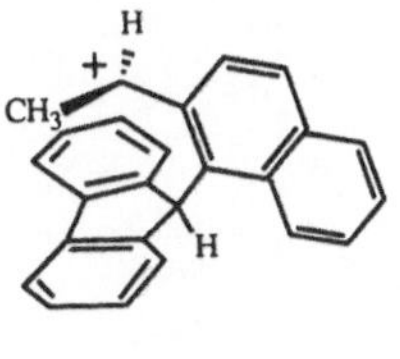

sp*(S*)-4-5 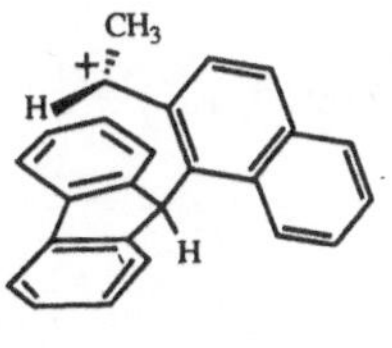   sp*(R*)-4-5

The results also imply that, if a cationic intermediate is formed over the fluorene ring, it should be stabilized by the participation of the π-electrons of the fluorene ring because of their proximity. Some addition reactions show effects of this kind.

Addition of bromine to the double bond in 9-(2-vinyl-1-naphthyl)fluorene (**4-3**) takes place normally and gives the corresponding dibromo compounds (**4-6**) for both the *ap* and the *sp* forms. No cyclization was observed even for the *sp* form. This means that a naked carbocation is not formed during the reaction but the cation is protected by the bromine participation, formation of the so-called bromonium ion, which is too unreactive with the fluorene ring to give a cyclized product. However, the relative rate of the bromine addition of the *ap* form is 0.8 when that of the *sp* form is taken as unity. This result can be ascribed to the participation of the π-system which enhances the reaction rates [3].

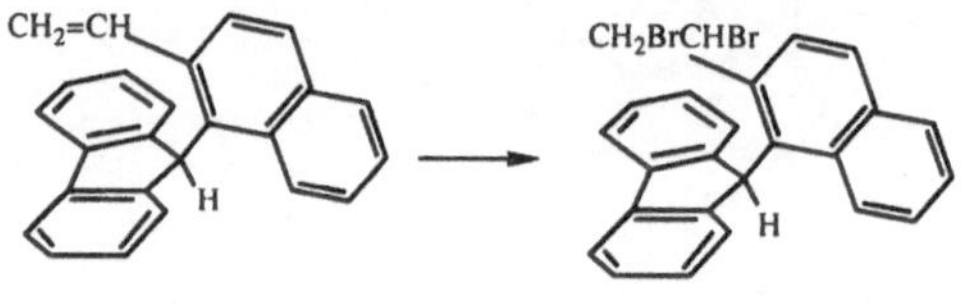

sp-**4-3** ap-**4-6**

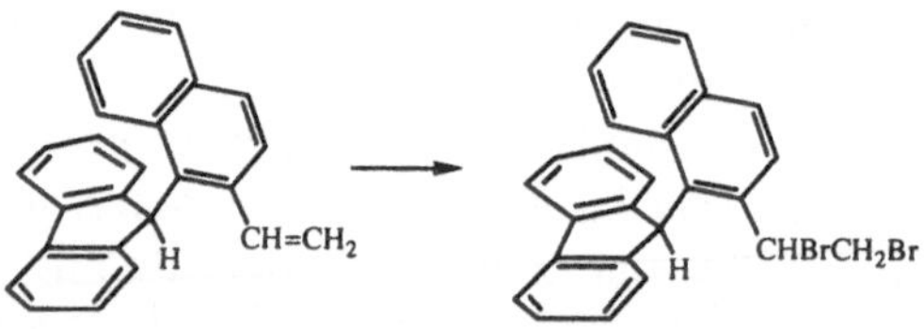

ap-**4-3** sp-**4-6**

By contrast, addition of chlorine to the same double bond gives a different result for the rotamers, when the products were examined. Addition of chlorine to *ap*-**4-3** gives a normal addition product, *sp*-[2-(1,2-dichloroethyl)-1-naphthyl]fluorene (*sp*-**4-7**), whereas the *sp*-isomer affords a substitution product (**4-8**), in addition to the adduct, in a 1.4 : 1.0 ratio [3]. This means that the chloride ion formed during the reaction attacks hydrogen at the β-position to

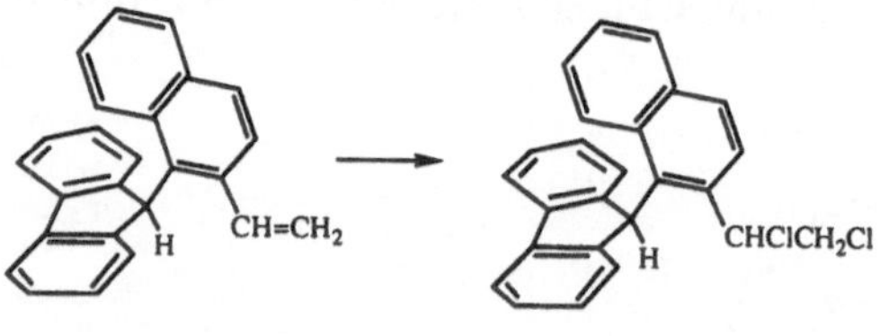

ap-**4-3** sp-**4-7**

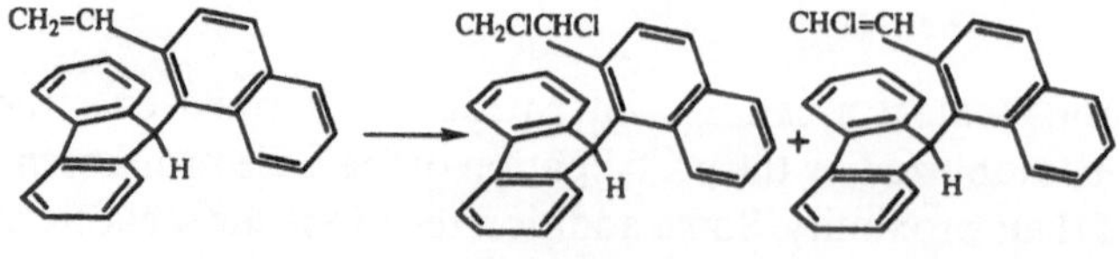

sp-**4-3** ap-**4-7** sp-**4-8**

remove a proton more easily than the carbocation. The difference with respect to the bromine addition is caused by the higher affinity of the chloride ion toward the proton. The chloro-olefin (4-8) originally obtained was a mixture of ca. 10:1 *trans* and *cis* isomers.

In these halogen additions to the olefinic bond, bridged cations should intervene, though the participation is rather weak. In the styrene derivatives, the bromine participation is known [4] to be very weak and practically no participation is observed in the case of chlorine [5]. However, if there is no chlorine-participation in the intermediate (4-9), we expect the cyclization to take place. Yet this reaction has not been observed. In addition to this fact, the *ap* and *sp* isomers react differently: if there is no participation of the chlorine, there is no basis to believe that the rotational isomers behave in different ways. Thus it can be assumed that even a chlorine substituent can participate to form a chloronium ion, though the participation is weak. Due to the π-participation in the chloronium ion formed from *sp*-4-3, the chloronium bridge is more unsymmetrical than that of the *ap*-isomer. Then the structure of the cation is such (4-9) that its plane is nearly parallel to the fluorene ring due to π-stabilization and the chlorine in the chloromethyl group is almost perpendicular to the fluorene due to the steric effect. Then the preferred formation of the *trans-sp*-4-8 can be rationalized by assuming deprotonation from the less hindered side of the almost-open cation (4-9).

If a cyclic transition state is required in an addition reaction, the reactivity of *ap*-4-3 could be different from that of the *sp* isomer because the steric environment of the reaction site is more congested in the former than in the latter. In order to see this kind of effects on the reactivities, various addition reactions which proceeds through a cyclic transition state/intermediate were examined. The results of competitive reactions are summarized in Table 4.1 [3].

The results in Table 4.1 indicate that the steric effect is rather small because both isomers exhibit almost the same reactivity, the highest difference being only 2.1 in terms of relative rates. These results are considered to be derived by two factors. One is the fact that the vinyl group has a very small hydrogen at its 1-position. The UV spectra of these compounds suggest indeed that the vinyl group is coplanar with the naphthalene ring both in the *ap* and the *sp* forms. In these conformations, the formation of the cyclic intermediate/transition state will be minimally affected by the steric effect of the fluorene ring. The other factor is the π-participation. Since the examples shown in Table 4.1 are in many cases the reactions between the olefin and an electron-demanding reagent, the

Table 4.1. Relative rates of addition reactions to the vinyl group in 9-(2-vinyl-1-naphthyl)fluorene (**4-3**) rotamers

Reagent (Solv)	Products	Relative Rates (k_{ap}/k_{sp})
Borane, THF	alcohols[a]	2.1
H$_2$/Pd, EtOH	ethyl compd	1.8
HN=NH,EtOH	ethyl compd	0.66
mCPBA, CH$_2$Cl$_2$	epoxide	0.81
O$_s$O$_4$,Et$_2$O	diol	0.90
C$_6$H$_5$ICl$_2$,CH$_2$Cl$_2$	dichloro compd	0.66

[a] After oxidation with alkaline hydrogen peroxide. The product was a 7:3 mixture of a 2-hydroxyethyl and 1-hydroxyethyl compounds.

intermediate/transition state will be stabilized by the π-participation which should enhance the reaction rates of sp-**4-3**.

In this respect, it is interesting to see the outcome of reactions of a cationic species with 9-[2-(1-methylethenyl)-1-naphthyl]fluorene (**4-10**). In the sp form of this compound, if the α-carbon atom of the 2-substituent in the naphthalene system is substituted by three non-hydrogen atoms, the steric effect is prohibitively strong for the existence. This means that a reaction to form such a 2-substituent must be very slow in the case of sp-**4-10**, whereas such a steric hindrance should not be so severe in the ap-**4-10**.

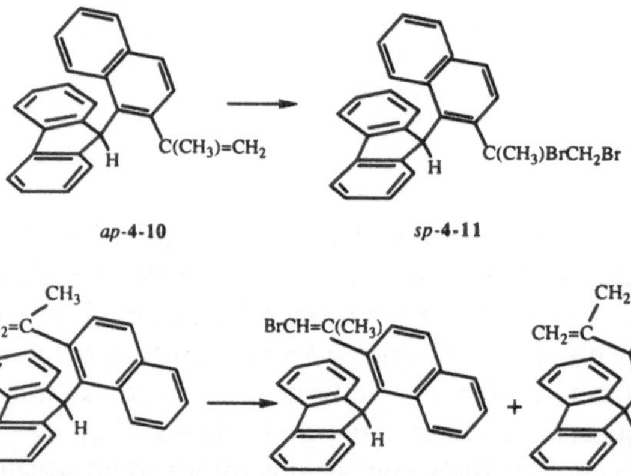

ap-**4-10**   sp-**4-11**

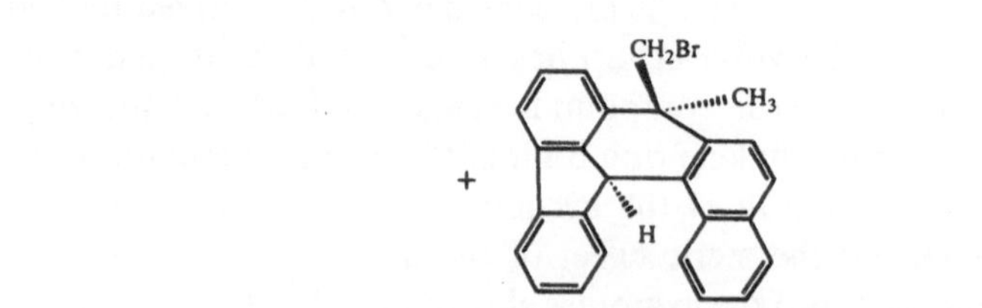

sp-**4-10**   ap-**4-12**   ap-**4-13**

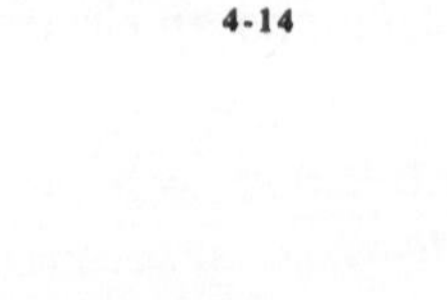

4-14

As an example of this class of reactions, addition reaction of bromine to **4-10** was examined [6]. The *ap*-form afforded a normal product, 9-[2-(1,2-dibromo-1-methylethyl)-1-naphthyl]fluorene (**4-11**), whereas the *sp* form gave no normal addition product. Instead, bromoolefins (**4-12** and **4-13**) and a cyclized product (**4-14**) were formed. Formation of the bromoolefins may be understood by considering the elimination of a proton from the intervening cation by a bromide ion. Formation of the cyclized compound is also a consequence of the reaction of the cationic intermediate with the fluorene ring.

Thus, the carbocation formed by addition of a bromine cation to the olefinic bond cannot be attacked at the positive carbon due to the steric effect. This type of olefin formation is exemplified by the addition of halogen to *trans*-cyclooctene [7] and crowded triptycene derivatives [8]. UV spectra of these compounds suggest that the olefinic moiety is not coplanar with the naphthalene nucleus in both forms and this structure is consistent with the molecular models. Once a cation is formed by addition of the bromine cation to the double bond, the rear side of the cation is blocked by the fluorene ring. Therefore normal addition reaction cannot be completed.

With this in mind, it is also interesting to see the fate of the similar carbocation formed from the *ap* form. Thus the cation was produced by the action of silver *p*-toluenesulfonate on the bromine adduct (**4-11**). This salt was used because the acidity of *p*-toluenesulfonic acid is comparable with hydrogen bromide [9] and consequently the proton affinity of the *p*-toluenesulfonate anion should be comparable with that of the bromide anion.

sp-**4-11**   sp-**4-12**   sp-**4-13**

The reaction gave bromoolefins (**4-12** and **4-13**), which were identical with those obtained from the *sp*-olefin (*sp*-**4-10**) except for their rotational positions: this was proved by thermal equilibration. Interesting is the formation ratio of **4-12** to **4-13**. **4-13** is more abundantly formed from the bromine adduct of the *ap* form. Examination of analogous compounds suggests that the bromonium ion participation in the *sp* cation is responsible for the difference because in the cation formed from *ap*-**4-10** and bromine the cation is more open in nature due to the π-participation.

Again no cyclization product is obtained from *sp*-**4-11** by debromo-ionization. This is due to the fact that the cation which intervenes cannot come to the top of the π-orbitals of the fluorene ring because of the structural requirement.

4.1.1.2 Addition to Carbonyl Groups

Addition reactions to a carbonyl double bond were also examined. 9-(2-Formyl-1-naphthyl)fluorene (**4-15**) is a mother compound. The formyl group in this compound is coplanar with the naphthalene ring in both the *ap* and the *sp* forms, as far as spectroscopic evidence indicates.

sp-**4-15**   ap-**4-15**

Table 4.2 [10] shows relative rates of the rotamers of the formyl compound (**4-15**), that would proceed by addition of a reagent to the carbonyl double bond. The rates are a little slower in the *ap* form than in the *sp* and contrast with those in the addition reactions to the olefinic double bonds. This is reasonable, because in the addition to olefins a cationic species reacts first and the transition state is stabilized by the π-participation, whereas in the addition to the carbonyl double bond an anionic species is added first and no significant stabilization by the π-system is expected. Thus the steric effects manifested over the fluorene ring are the main factor that retards the reaction in the *ap* form.

Grignard addition to the formyl group of compound **4-15** gives the relative rate, k_{sp}/k_{ap}, of ca. 1.8, whereas that of oxime formation and that for sodium borohydride reduction are much larger. Single electron transfer, which is now assumed to be an initial step of the Grignard addition, may be responsible for the difference from other reactions.

Addition to a double bond of other carbonyl groups is quite a different story, as it was for the addition to the olefinic double bond of **4-10**, because addition of an anionic species to a carbonyl group of a ketone, for example, produces an α-carbon which carries three non-hydrogen atoms. This is sterically very demanding and formation of such a carbon atom should be very slow. These considerations lead to a prediction that, whereas the carbonyl groups in the *sp* compound (*sp*-**4-16**) normally react with various reagents, those in the *ap* compounds (*ap*-**4-16**) will refuse to react. Indeed, such examples were found

Table 4.2. Relative rates of addition to the formyl group in 9-(2-formyl-1-naphthyl)fluorene (**4-15**) rotamers

Reaction	Relative Rates (k_{sp}/k_{ap})
Oxime formation	3.7
Reduction with NaBH$_4$	7.6
Addition of C$_6$H$_5$MgBr	1.8
Addition of CH$_3$MgI	1.7

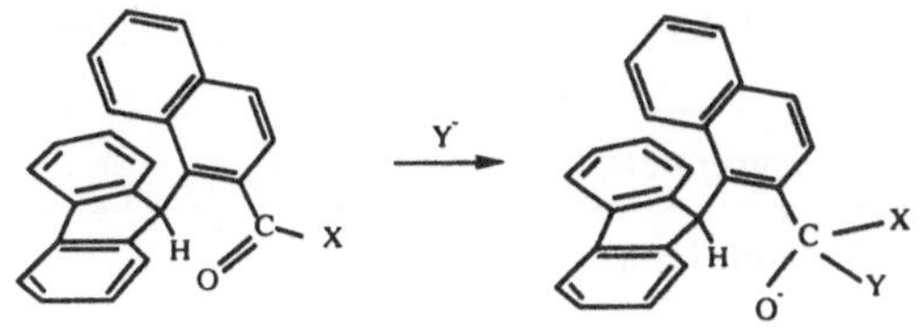

sp-4-16 → *sp*-4-17

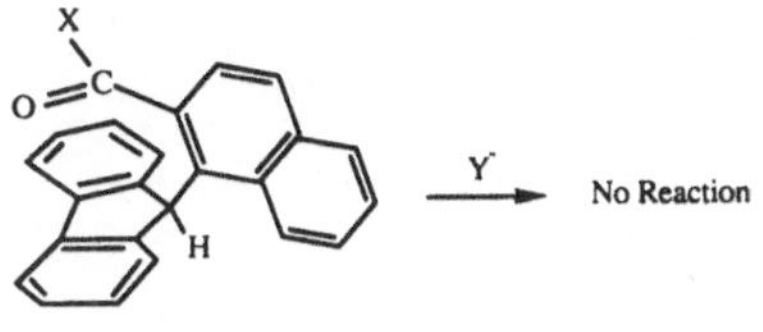

ap-4-16 → No Reaction

in Grignard addition to a ketone (**4-16**: X=CH$_3$), hydrolysis of an imidazolide (**4-16**: X=1-imidazolyl), and sodium borohydride reduction of a ketone (**4-16**: X=CH$_3$): the *sp* forms of these compounds reacted smoothly to afford *sp*-**4-17**, whereas the *ap* forms were recovered under the reaction conditions for all the compounds examined [11].

4.1.1.3 Elimination Reactions

Some elimination reactions of vicinal dihalogen compounds, and others which are related, to produce olefins are known to be of E2 type. The results of an example of such reactions, elimination from compound **4-18** to produce compound **4-3** are summarized in Table 4.3 [12]. Elimination of bromine by a metal gives rather a small relative rate, but others give rather large values. This is indicative that the elimination by the metal is different from others in the mechanisms of the reaction. This probably proceeds via multisteps. The slow elimination of bromine from the *ap* forms relative to the *sp* is attributed to the steric effect in the transition state of the E2 elimination. Elimination of a silyl

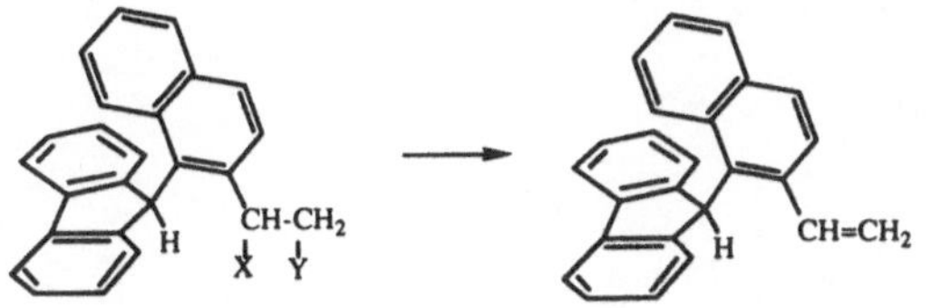

sp-4-18 → *ap*-4-3

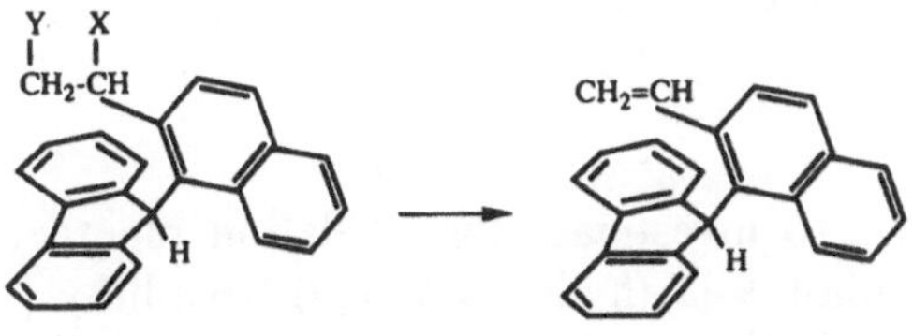

ap-4-18 → *sp*-4-3

Table 4.3. Relative rates of elimination reactions in 9-[2-(1,2-disubstituted ethyl)-1-naphthyl]-fluorene (**4-18**) rotamers

X	Y	Reagent	Temperature/°C	k_{sp}/k_{ap}
Br	Br	Zn	Room Temp	1.4
Br	Br	LiAlH$_4$	-15	8.6
Br	Br	LiAlH$_4$	0	8.8
Br	Br	LiAlH$_4$	40	8.7
Br	Br	Na$_2$Te	Room Temp	3.1
OH	Si(CH$_3$)$_3$	H$_2$SO$_4$	15	2.6
OH	Si(CH$_3$)$_3$	H$_2$SO$_4$	0	2.5

group and an oxygen-containing group in the vicinal position is known as the Peterson reaction [13] and in the presence of acid proceeds in *trans* fashion. The results shown in Table 4.3 give support to the concept that the reaction takes place via simultaneous elimination of the two groups.

A large steric effect provided by the fluorene system in the rotamers of 9-arylfluorene was first noticed in the oxidation of the formyl compound (**4-15**). Oxidation of the aldehyde to the corresponding carboxylic acid is known to proceed via addition of water to the carbonyl group, formation of a chromate ester, and the elimination of the low oxidation state of chromium as an ester [14]. The rate-limiting step is the last step, though whether the elimination is intramolecular or intermolecular is not known. The finding that the *sp* aldehyde was oxidized normally to afford a carboxylic acid (**4-19**) but the *ap* form was practically not oxidized by chromium(VI) oxide at room temperature, was a great surprise [10]. The *ap* aldehyde gave a lactone (**4-20**) by such an oxidation when the reaction was allowed to proceed for a long time.

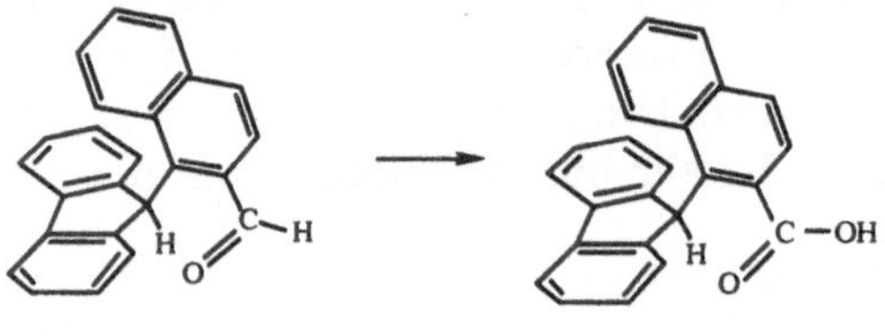

sp-**4-15**   *sp*-**4-19**

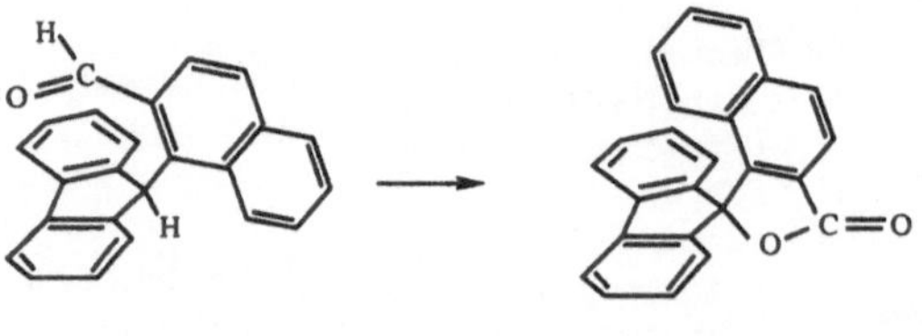

ap-**4-15**   **4-20**

The finding prompted the authors to investigate the oxidation of other alcohols. Oxidation of a primary alcohol, 9-[2-(hydroxymethyl)-1-naphthyl]-fluorene (**4-21**: R=H) proceeds smoothly both in *ap* and *sp* forms with a relative rate of 0.67 in favor of the *ap*. However, the relative rates, k_{sp}/k_{ap}, for the oxidation of secondary alcohols (**4-21**) were ca. 34 and ca. 27, respectively, for the R of a phenyl and a methyl [10].

The very large retardation of oxidation in the secondary alcohols in the *ap* forms is the model for the oxidation of the aldehyde and the slow oxidation of the *ap* aldehyde can now be attributed to the steric effects given by the fluorene ring in the last elimination step. The formation of the lactone (**4-20**) is ascribed to a fairly easy oxidation of the fluorene group at the 9-position, though it is not clear whether the oxidation at the position takes place prior to or after the oxidation of the formyl group.

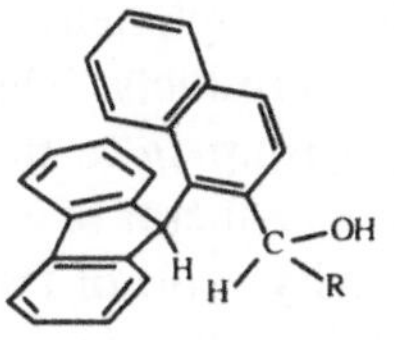

sp-4-21 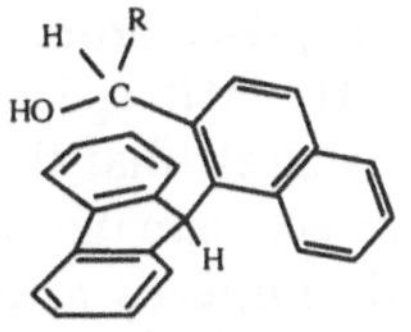   ap-4-21

Other elimination reactions of rotamers are known in the case of oxime-dehydration. The dehydration of oximes is known to proceed in *trans* fashion [15] and thus strong steric effects are expected in the *ap* form. Indeed, dehydration of sp-9-(2-formyl-1-naphthyl)fluorene oxime (**4-22**) to form sp-9-(2-cyano-1-naphthyl)fluorene (**4-23**) proceeded much more easily than that of the *ap*-counterpart, the relative rate, k_{sp}/k_{ap}, being ca. 18 [10].

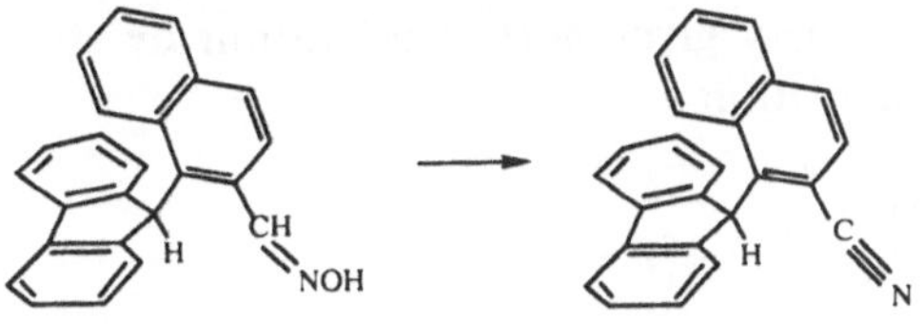

sp-**4-22**   sp-**4-23**

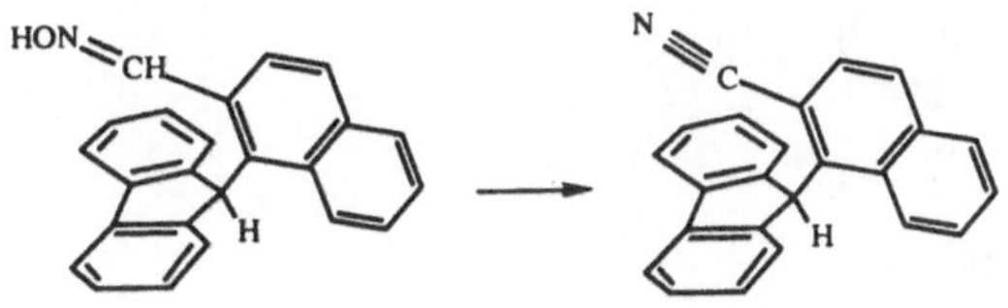

ap-**4-22**   ap-**4-23**

4.1.2 Substitution Reactions

Two methyl groups in 9-(2,6-dimethylphenyl)fluorene (**4-24**) are diastereotopic because rotation about the C_{ph}–C_9 bond is now frozen: substitution of a substituent for one of the hydrogens in the methyl group which is *ap* to the

4-24   sp-4-25   ap-4-25

C_9–H gives a diastereomer of that formed by substitution of the methyl group *sp* to the C_9–H. This type of difference in the reactivity is usually neglected but should be seen if the rotation is frozen.

Bromination of compound **4-24** with *N*-bromosuccinimide in boiling benzene gives a 2.2:1 mixture of *sp* and *ap*-9-(2-bromomethyl-6-methylphenyl)-fluorene (**4-25**) as judged from the ^{1}H NMR spectra of the products. The isolated ratio was 17:6. These results suggest that the methyl group *sp* to the C_9–H bond is more vulnerable to the attack than the *ap* methyl by a factor of ca. 2.5. However, this value cannot be taken directly as the reactivity ratio of the methyl groups, because slow rotation about the C_{ph}–C_9 bond takes place under these conditions. The equilibrium constant for the *ap* → *sp* process was found to be 3.6 ± 0.2 for **4-25**. Thus to a small extent, the *sp* form must have been formed by rotation. The *sp* methyl group is concluded to be more reactive than the *ap* by a factor of ca. 2 under these conditions [16]. The tendency is believed to be the consequence of the steric effects that protect the *ap*-methyl as well as the π-participation which enhances the reactivity of the *ap*-methyl.

The rotational isomers of 9-(2-bromomethyl-6-methylphenyl)fluorene (**4-25**) exhibit interesting differences in various substitution reactions. This is because, at the *ap* site, the fluorene moiety gives both deceleration by steric hindrance and acceleration by π-participation.

4-25   4-26

When compound **4-25** was treated with pyridine, it gave the corresponding pyridinium salts (**4-26**) as a result of a normal Menshutkin reaction [17]. The relative rate, k_{sp}/k_{ap}, was 22. Reaction with 2-methylpyridine gave a still higher relative rate, 34. These results indicate that in the S_N2 reactions the steric effect of the fluorene ring is very important and retards the reaction of the bromomethyl group in the *ap* position relative to C_9–H.

This steric effect suffices to cause methanolysis of practically only one of the two bromomethyl groups in 9-[2,6-bis(bromomethyl)phenyl]fluorene (**4-27**).

The methanolyzed group is *sp* to the C_9–H bond and the product is *ap*-9-[2-(bromomethyl)-6-(methoxymethyl)phenyl]fluorene (**4-28**). This is another example of the difference in the reactivity, chemoselectivity, of a diastereotopic pair of groups and a high selectivity results because of the steric effect.

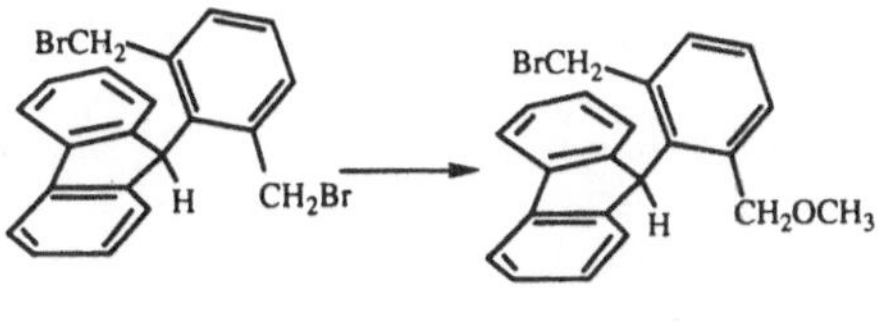

4-27   *ap*-**4-28**

When methoxide is used as a nucleophile, there is competition between the substitution of the bromo in the bromomethyl group and abstraction of a proton from the 9-position of the fluorene, because the methoxide is a strong and small base. In *ap*-**4-25**, the S_N2 reaction at the bromomethyl group is so slow that the deprotonation from the 9-position wins the competition and, in this sodio compound or anion, internal rotation is much easier than in the 9–H compound. Thus in *ap*-**4-25**, the major product is a spiro compound (**4-30**) as a result of an intramolecular S_N2 reaction of the 9-anion with the bromomethyl group in its proximity and a trace of the corresponding methoxy compound (*sp*-**4-29**) is isolated together with a few percent of the isomeric methoxy compound (*ap*-**4-29**). By contrast, the major product from *sp*-**4-25** is *sp*-**4-29** together with a small amount of the spiro compound (**4-30**) and the isomerized one (*ap*-**4-29**). If *tert*-butoxide, which is known to be a nucleophile that reacts sluggishly in the S_N2 type reactions, was used, the sole product was compound **4-30** [17].

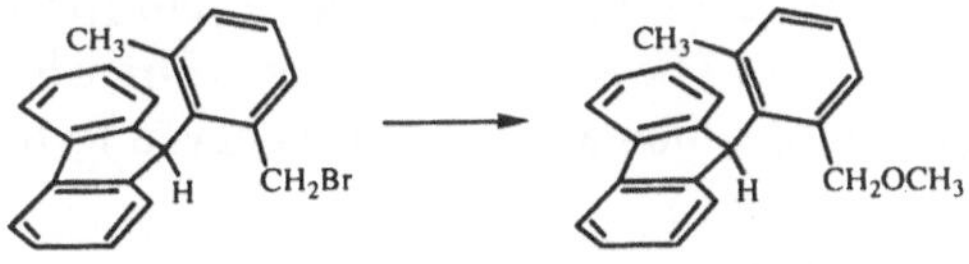

sp-**4-25**   *sp*-**4-29**

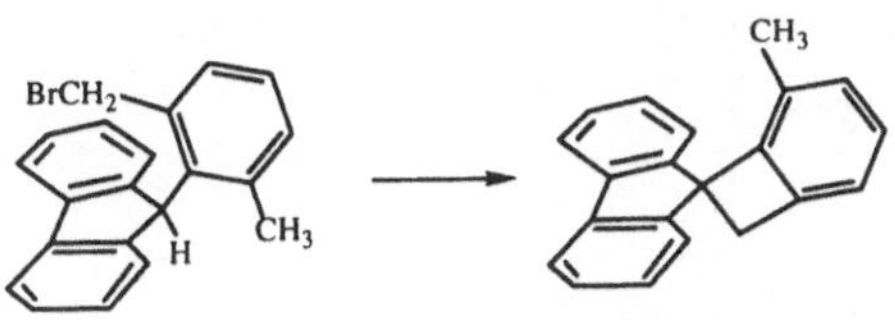

ap-**4-25**   **4-30**

The difference in the reactivity of 9-[2-(bromomethyl)-6-methylphenyl]-fluorene (**4-25**) is more dramatic, when it is heated in ca. 6:1 trifluoroacetic acid and chloroform [18]. The *ap* form afforded a cyclized product (**4-31**), whereas the *sp* form is recovered unreacted after heating for 40 h at 62 °C. The formation of the cyclized product is interpreted as a result of formation of a cation by

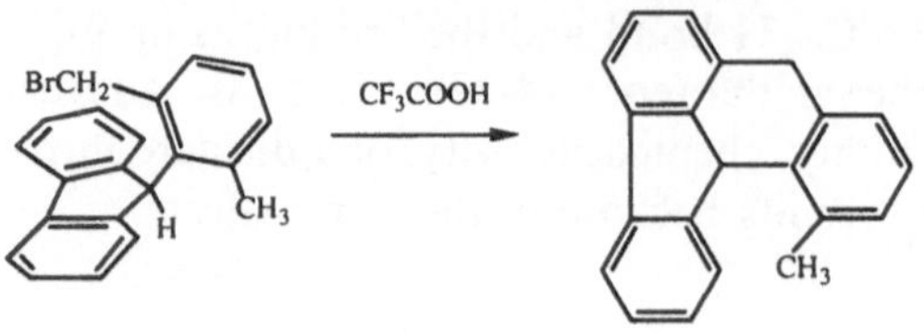

ap-**4-25**

4-31

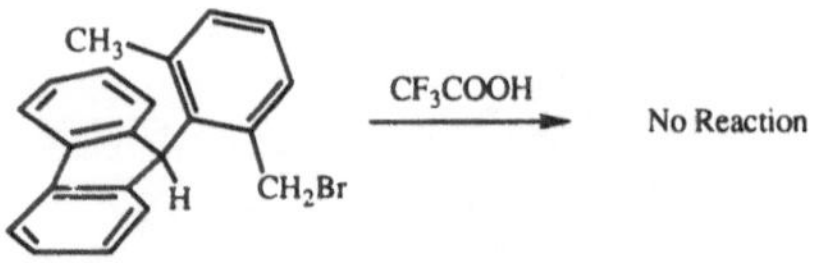

sp-**4-25**

heating in a polar medium followed by Friedel-Crafts type alkylation. To obtain the cyclized product in a reasonable yield, the concentration of the solution should be rather low, because polymerization takes place at high concentrations. This is because, for the cyclization, internal rotation, which requires rather high energy, has to take place for the intramolecular reaction. An intermolecular Friedel-Crafts reaction becomes much easier in these cases.

However, the above results should not be taken to mean that the *sp*-**4-25** does not ionize under the conditions. In trifluoroacetic acid which contains 5% chloroform and 3% methanol, the *sp* form gives the corresponding methoxy compound (*sp*-**4-29**) in a good yield, whereas the *ap* isomer afforded the cyclized compound (**4-31**) under the same conditions. The relative rate of disappearance, k_{ap}/k_{sp}, is 5.6. The results suggest that the ionization takes place under these conditions even in the *sp* form. The ion pair cannot react to give the products in trifluoroacetic acid, because the solvent molecule is one of the least nucleophilic reagents and the fluorene ring is not in the favorable position for cyclization. The ion pair returns to the covalent species. The same phenomena are also observed in 1,1,1,3,3,3-hexafluoro-2-propanol which at present is believed to be the least nucleophilic [19].

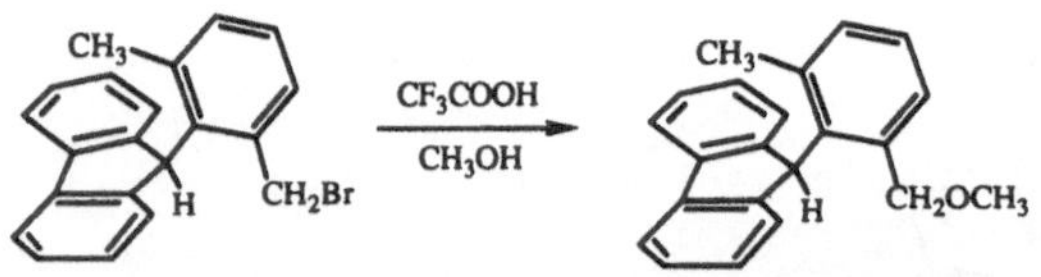

sp-**4-25**   *sp*-**4-29**

The formation of the cyclic compound (**4-31**) in the case of the *ap* form is interpreted to mean that the reaction of the cation with the fluorene ring won the competition with the reaction with methanol. This result may be derived from the close proximity of the benzene ring in the fluorene moiety or the low concentration of methanol in the system, especially because the conditions are protonating, though the cation should react with the solvent under normal conditions.

Table 4.4. Relative rates in silver salt assisted ionization of **4-25**

Silver Salt	Solvent[a]	Temp/°C	Reaction Period/min	k_{sp}/k_{ap}
AgNO$_3$	CH$_3$CN	25	40	5.9
AgClO$_4$	CH$_3$CN	60	1200	1.3
AgClO$_4$	C$_6$H$_6$	22	120	1.6
AgClO$_4$	HFP[b]	14	72	0.78
AgClO$_4$	TFA[c]	25	13	0.94

[a] These solvents contain ca. 5% (v/v) chloroform,
[b] 1,1,1,3,3,3-hexafluoro-2-propanol,
[c] Trifluoroacetic acid.

In order to examine this point, cations were generated from compound **4-25** by various silver salts in various solvents. The results are summarized in Table 4.4. The products are those which are derived by the reaction of the cations with the solvent, except for the case of silver nitrate, when cations react with the nitrate anion instead of the solvent molecules.

Interesting is the fact that the *sp* isomer reacts more easily than the *ap*, though the factor is not very remarkable. This is in sharp contrast to the behavior in trifluoroacetic acid and in hexafluoro-2-propanol. Furthermore, solvolysis of the bromomethyl compound (**4-25**) in formic acid, which is known to be close to S$_N$1, proceeds smoothly and both the *sp* and the *ap* forms give formates to indicate that the nucleophilicity of formic acid wins the competition with the reactivity of the fluorene ring. The relative rate, k_{sp}/k_{ap}, of the formolysis was 1.6 at 42 °C.

These results are interesting in that, in very poorly nucleophilic solvents, trifluoroacetic acid and hexafluoro-2-propanol, the *ap* form is more reactive than the *sp*, whereas the reverse is true in other solvents which are more nucleophilic than those of the polyfluoro compounds. This implies that the *ap* cation is more stable than the *sp* in very poorly nucleophilic solvents but vice versa in nucleophilic solvents.

sp-4-32   ap-4-32

The interpretation of this phenomenon is given by the taking advantage of the solvation energy of the cation. In the *ap* cation, at least one side of it is blocked by the fluorene ring. Thus the solvation shell will be completed with participation of the fluorene ring. In contrast, in the *sp* cation, the solvation shell may be formed by solvent molecules only. This statement could be an exaggeration, but it is still true that a number of the solvent molecules that participate in forming the solvation shell of the *sp* cation is more than that for the *ap* cation. Thus the solvation shell may be modeled as shown by **4-32** where S represents solvent molecules.

Then, if solvent molecules solvate more efficiently than the solvation including the fluorene ring, the *sp*-**4-25** is more vulnerable toward ionization than *ap*-**4-25**. If the solvation by solvent molecules is less effective than that including the fluorene ring, the *ap* form reacts more easily than the *sp*. It deserves mentioning that the difference in reactivities of rotational isomers provide information on solvation.

The reaction of 9-[2-(1-hydroxyethyl)-1-naphthyl]fluorene (**4-33**) with thionyl chloride is another example which shows interesting differences with different rotamers [1]. Its *sp* form gives a normal product, *sp*-9-[2-(1-chloroethyl)-1-naphthyl]fluorene (**4-34**) in every solvent examined. The *ap* form gives a dehydration product, *sp*-9-(2-vinyl-1-naphthyl)fluorene (**4-3**), in addition to the corresponding *ap*-chloride (**4-34**). The formation of the olefin is sometimes observed in aliphatic secondary alcohols when it is treated with thionyl chloride but has not been found in benzyl-type secondary alcohols.

The formation ratio of the olefin (**4-3**) to the chloride (**4-34**) was dependent on the solvent used. As summarized in Table 4.5, the formation of the chloride is

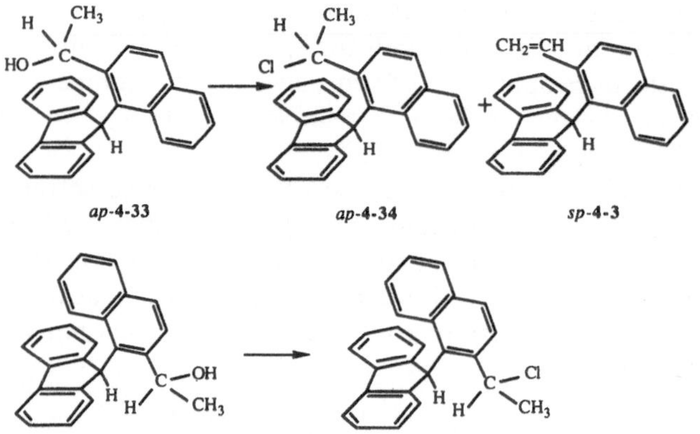

ap-4-33   ap-4-34   sp-4-3

sp-4-33   ap-4-34

Table 4.5. Dehydration vs. chlorination in the reaction of compound **4-33** with thionyl chloride

Solvent	Dielectric Constant[a]	From *ap*-alcohol	From *sp*-alcohol
		(*sp*-**4-3**/*ap*-**4-34**)	(*ap*-**4-3**/*sp*-**4-34**)
hexane	1.88	71/29	< 5/ > 95
dioxane	2.21	67/33	< 5/ > 95
CH_2Cl_2	8.93	60/40	< 5/ > 95
$SOCl_2$	9.25	38/62	< 5/ > 95
CH_3NO_2	35.9	20/80	< 5/ > 95

[a] Dielectric constants of solvents.

favored in polar solvents but the olefin is the major product in nonpolar solvents. Substitution of three deuteriums for three hydrogens in the methyl group of *ap*-**4-33** reduced the formation of the olefin significantly.

An interpretation which accommodates all of these findings as well as the reaction mechanisms which had been established [20] is as follows. Thionyl chloride reacts with the alcohol (**4-33**) to form a chlorosulfinate which then decomposes to an ion pair and sulfur dioxide. During this process, the molecule has to pass a cyclic transition state which is space-demanding and energy-demanding in the *ap* position. Since the transition state (**4-36**) is not congested in the *sp* position, a normal reaction takes place to produce the chloride. In the *ap*, then the olefin formation becomes competitive with the chloride formation, because for the former reaction steric hindrance is less severe due to the fact that one of the hydrogens of the methyl group is always far enough away from the fluorene ring to make the transition state (**4-37**) energy low relative to that (**4-36**) for the chloride formation. The solvent effect can be interpreted to mean that the polar solvent favors loose ion-pairs to make the cyclic transition state earlier and thus relieving the steric strain in the transition state (**4-36**) to allow formation of the chloride.

4-36   **4-37**

4.1.3 Acid–Base Properties

Acidity of a carboxylic acid in a crowded environment is known to be lower than that in a less crowded position. An example is cyclohexanecarboxylic acid, the equatorial isomer is more acidic than the axial counterpart by ca. 1.0 pKa unit [21]. This is interpreted by the taking advantage of the solvation energy. The solvation of the crowded carboxylate anion is less favorable than that of the less crowded carboxylate.

4.1.3.1 Acids and Bases in Substituents of the Aryl Group

Similar situations due to the solvation for crowded and less crowded sites are expected for *ap* and *sp* carboxylic acids in the 9-arylfluorene series: the *ap* form is less acidic than the *sp* because the carboxylate anion in the *ap* position will be less stabilized by solvation than the *sp*. This expectation is realized in *ap* and *sp*-1-(9-fluorenyl)-2-naphthoic acid (**4-38**), pKa being 7.1 and 6.1 in 4:1 ethanol-water at 25 °C, respectively [22].

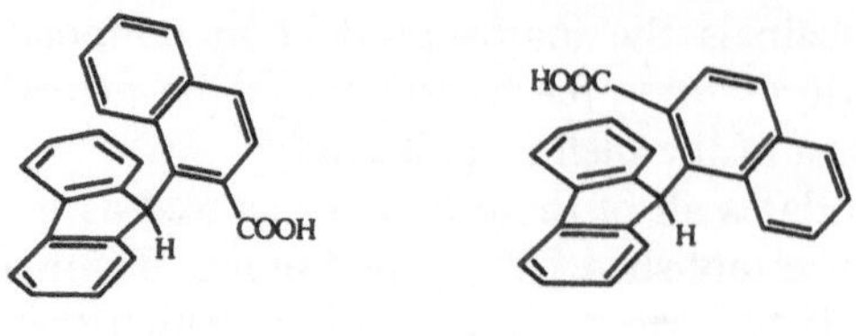

sp-4-38   ap-4-38

On the basis of this concept, acidities of conjugate acids derived from amines of the 9-arylfluorene series are sometimes anomalous. In Table 4.6 are summarized the pKa values of the conjugate acids of methylated and unmethylated 9-(2-amino-1-naphthyl)fluorene rotamers (**4-39**) [22]. The data suggest that, whereas amines that carry hydrogen(s) before forming the triflates are normal because the *ap* amines are weaker bases than the *sp*, the dimethylamino compounds are anomalous because the *ap* form is more basic than the *sp*. This puzzling phenomenon is explained by the presence of NH-π interactions in the *ap* form of the ammonium salt and has been studied in detail by taking other compounds as shown below.

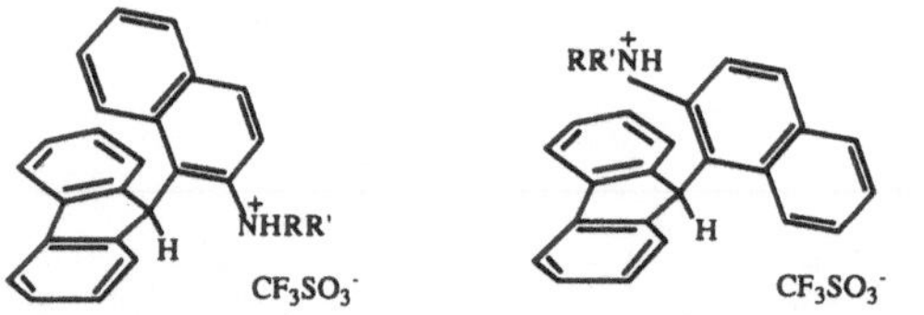

sp-4-39   ap-4-39

Since the anilines are weak bases, it should be more convenient to use aliphatic amines for the study. Rotamers of 9-(2-dimethylaminomethyl-6-methylphenyl)fluorene (**4-40**) were submitted to investigation. The pKa' values for the trifluoroacetates of the *ap* and *sp* rotamers were 6.45 and 5.85, respectively, in 3:2 tetrahydrofuran-water at 25 °C [23].

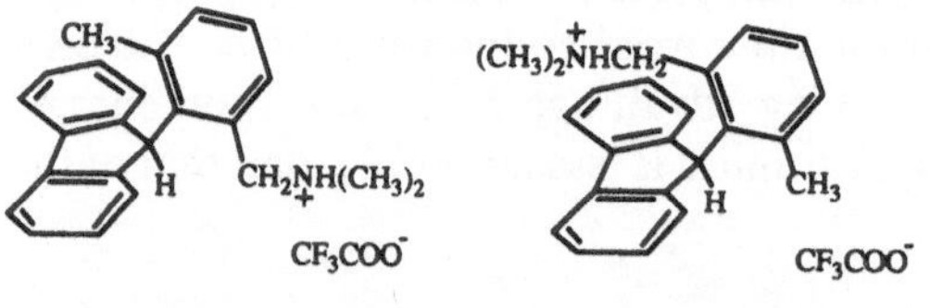

sp-4-40   ap-4-40

Table 4.6. pKa values of the triflate salts of 9-[2-(alkylamino)-1-naphthyl]fluorene (**4-39**) rotamers in 4:1 ethanol-water

R	R'	pKa(sp)	pKa(ap)	ΔpKa
H	H	3.3	2.6	− 0.5
H	CH$_3$	3.3	2.7	− 0.6
CH$_3$	CH$_3$	2.7	2.8	0.1

^{1}H and ^{13}C NMR spectra of these isomers indicate that the methyl groups on the nitrogen in the *ap* form are located in the region where ring current effect of the fluorene moiety is significant. The NH-π interactions are considered to be stronger when the electronegativity of the NH group becomes high, because it is the case for the OH-π interactions as studied by infrared spectroscopy [24]. There is a bigger chance for the ammonium NH group to form the NH-π hydrogen bond than the normal NH group without the positive charge on the nitrogen. Then it is reasonable to assume that the NH-π interactions, which are not possible in the *sp* but are possible in the *ap*, are responsible for the observed phenomenon. The NH-π interacting structure can be visualized as that in **4-41**.

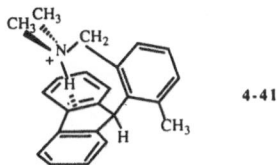

4-41

If the NH-π interactions are important, the tetrahydrofuran-water system is not a good one to see a big difference in the basicities of the amines, because both tetrahydrofuran and water are proton accepting in hydrogen-bond formation. The NH-π hydrogen bond is competing with the NH–O hydrogen bond in the system and the former is normally weaker than the latter. If a solvent, which does not accept hydrogen in hydrogen bond formation, is used, the difference should be enhanced.

However, it is not easy to determine dissociation constants in nonpolar solvents because conductometric titration is not possible for those solvents. Although the absolute value for the dissociation of ammonium salts in nonpolar solvents cannot be accessed, the difference in the dissociation constants can be determined by another means.

Dissociation constants of the rotamers can be defined in the following equations.

$$K_{sp} = \frac{[sp\text{-amine}][CF_3COOH]}{[sp\text{-salt}]} \tag{1}$$

$$K_{ap} = \frac{[ap\text{-amine}][CF_3COOH]}{[ap\text{-salt}]} \tag{2}$$

Dividing Eq. 1 by Eq. 2 gives following relationship.

$$\frac{K_{sp}}{K_{ap}} = \frac{K[\text{salt}]}{K[\text{amine}]} \tag{3}$$

Thus ΔpKa can be determined by measuring the population ratios of the rotameric free amine and the salt. This is possible by means of NMR spectra.

In chloroform-*d*, rotamer population ratios of the free amine and the salt, *ap/sp*, were found to be 1/3.66 and 9.70, respectively. Thus the ΔpKa amounts to

1.55 [23]. This large difference in the pKa values of the rotamers gives support to the idea that the NH-π interactions are responsible for the anomalously strong basicity of the *ap* amine with respect to the *sp*. Because the solvation of an ammonium salt by a nonpolar solvent is poor, the ammonio group in the *sp* position is less stabilized than that in the *ap* in which π-electrons participate to stabilize the cation.

Since the solvation energy of ammonium salts derived from tertiary amines should be affected by the chain lengths of the alkyl groups attached to the nitrogen, it is worthwhile examining the difference in basicities of amines (**4-42**) with various alkyl groups. Table 4.7 lists the results of such study on aqueous tetrahydrofuran solutions [23].

sp-**4-42**   ap-**4-42**

The difference in pKa's of these amines is ca. 1.0, when the alkyl chains becomes longer than a methyl one in tetrahydrofuran-water. Especially noteworthy is the fact that the pKa's of both the *sp* and *ap* forms generally decrease, as the chain lengths become longer, and the effect is more significant for the *sp* forms. This is to be expected because the stabilization by solvation is more effective in the *sp* form and that should be affected by the chain lengths.

Solvent effects on the differences of the pKa values of the trifluoroacetate salts of these amines are summarized in Table 4.8 [23]. Toluene and chloroform give the largest differences. This is easy to understand because the hydrogen

Table 4.7. pKa values of trifluoroacetate salts of 9-[2-(dialkylaminomethyl)-6-methylphenyl]fluorene (**4-42**) rotamers in 3:2 THF-water

R	R'	pKa(sp)	pKa(ap)	ΔpKa
CH_3	CH_3	5.90	6.45	0.55
C_2H_5	C_2H_5	5.70	6.80	1.10
C_2H_5	n-C_3H_7	5.22	6.15	0.93
n-C_3H_7	n-C_3H_7	4.98	6.12	1.14

Table 4.8. Solvent effects on the difference of pKa values of the salts (**4-42**)

R	R'	Toluene-d_8	THF	CH_3CN	CH_3NO_2	$CDCl_3$
CH_3	CH_3	1.33	0.83	0.83	1.16	1.55
C_2H_5	C_2H_5	1.52	1.14	1.08	1.18	1.53
C_2H_5	n-C_3H_7	1.53	1.05	1.18	1.23	1.54
n-C_3H_7	n-C_3H_7	1.65	0.96	1.12	1.13	1.68

accepting power of the atoms in these molecules is low. In contrast, tetrahydro-furan and acetonitrile give the smallest difference, although this is larger than in aqueous tetrahydrofuran. This can be attributed to the hydrogen accepting power of oxygen and nitrogen in the hydrogen-bond formation. Nitro groups are known to form very weak hydrogen bonds and this property is reflected in the intermediate ΔpKa values observed for the nitromethane solution.

The difference in basicities of the rotameric amines can also be observed in the dynamic behavior of the pair of compounds **4-42** ($R=R'=C_2H_5$) [25]. The behavior is the proton exchange between the amino-nitrogen and the trifluoro-acetate anion. This process can be studied by the dynamic NMR technique. In $CDCl_3$, the *sp* form shows faster proton exchange than the *ap*, indicating that *ap*-**4-42** ($R=R'=C_2H_5$) is a stronger base than the *sp*. Free energies of activation for the exchange were 14.1 and 16.4 kcal/mol at 298 K for *sp* and *ap* salts, respectively. By contrast, the same exchange reaction in tetrahydrofuran-d_8 proceeded with similar free energy of activation for the two isomers, though the *sp*-form gives slightly larger rates than the *ap*: 13.4 and 13.7 kcal/mol at 298 K for *sp* and *ap*-**4-42** ($R=R'=C_2H_5$), respectively.

Coming back to the problem of basicities of 9-(2-dimethylamino-1-naphthyl)fluorene (**4-39**: $R=R'=CH_3$) rotamers, one notices that the difference in basicities of the rotamers is diminished with respect to rotamers of the dimethyl-aminomethyl compound (**4-42**). This can be attributed to the unfavorable geometry for the formation of the NH-π hydrogen bond in the *ap* form due to the fact that the amino group is directly attached to the aromatic ring. When more than one hydrogen is present in the ammonium salt, hydrogen bonding both to the fluorene ring and to the solvent molecule is possible. This is the reason why the conjugate acids of the primary amine and the secondary amine show normal behavior, i.e. the *ap* form is a weaker base than the *sp*.

As expected, the difference in pKa's of these amines (**4-39**), i.e. those that bear at least one hydrogen atom before protonation, in chloroform, as studied by the population method described above, shows that the *sp* forms of the primary and secondary amines are stronger bases than the *ap* by ca. 1.0 pKa unit. However, the *ap*-form of 9-(2-dimethylamino-1-naphthyl)fluorene (free amine correspond-ing to **4-39**: $R=R'=CH_3$) is a stronger base than the *sp* by 1.3 pKa units.

Unfortunately, phenols of this series are too weakly acidic for us to measure their acidity precisely by the titration method in solvents in which they are soluble. The rough estimate indicates that the *ap*-9-(2-hydroxy-1-naphthyl)-fluorene (**4-33**: X=O) is an acid which is ca. 1.0 pKa unit weaker than its *sp* counterpart. pKa's of *sp* and *ap*-1-(9-fluorenyl)-2-naphthalenethiol (**4-33**: X=S)

sp-**4-43**   *ap*-**4-43**

in 4 : 1 ethanol-water are 8.6 and 10.6, respectively. Again the *ap* form is a weaker acid than the *sp* due to the steric effect. The enhanced difference in acidities of the thiols with respect to carboxylic acids may be attributed partially to the presence of weak SH-π interactions [22].

4.1.3.2 Deprotonation at the 9-Position of Fluorene

The 9-CH group of fluorene is acidic because the anion formed by deprotonation is stabilized due to being a 6π-electron system. Indeed, there is a wealth of literature which reports the properties of ions and ion-pairs of this system as typical organic anions. Therefore, deprotonation is another reaction which is worth examining as acid-base properties of the 9-arylfluorene system.

Deprotonation of *ap* and *sp*-9-(2-methyl-1-naphthyl)fluorene (**4-44**) was carried out with the use of butyllithium in benzene-hexane. It was found that the *sp* form reacts more easily than the *ap* by a factor of ca. 7 [26]. Since the 9-H in the *sp* is close to the naphthalene ring and that in the *ap* to the methyl group, the results can be taken as a reflection of the steric hindrance to the approach of the aggregated butyllithium to the reaction site: the van der Waal's radius of the methyl group is larger than the half-thickness of the π-system. Since both substrates and the solvents are hydrocarbons, butyllithium exists as a hexamer [27] and sterically demanding when it approaches a reaction site.

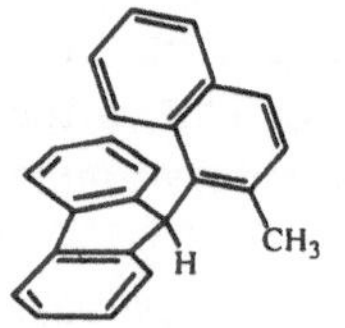

ap-**4-44** 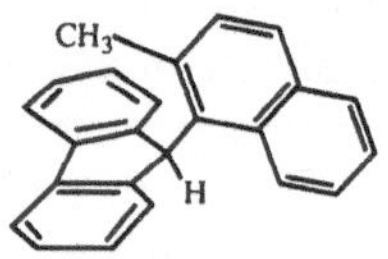   *sp*-**4-44**

On the stereochemical basis, as was found in the case of compound **4-44** rotamers, the lithiation of *sp*-9-(2-methoxy-1-naphthyl)fluorene (**4-45**) will be faster than that of the *ap* isomer by a factor of 10–100, because an oxygen atom is a little smaller than the π-system. Therefore, it was a surprise for initial investigators to find that deprotonation of the *sp* isomer was more than 1000 times faster than that of the *ap* isomer in hydrocarbon solvents [28].

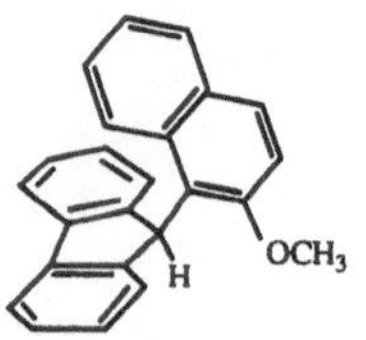

sp-**4-45** 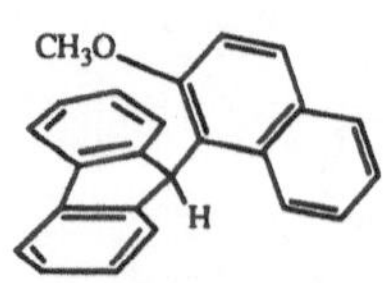   *ap*-**4-45**

Incidentally, it seems that internal rotation takes place after lithiation under these conditions, because the same *sp*-9-(2-methoxy-1-naphthyl)fluorene is recovered after protonation of the lithiated product. This is attributed to the lowering of the barrier to rotation in the lithiated compounds (**4-46**) and the oxygen-coordinated form of the *sp* isomer is much more stable than others such as the merely lithiated form of the *ap*.

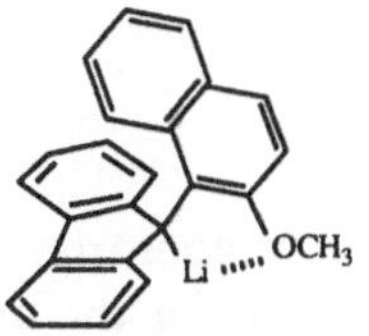

sp-**4-46** 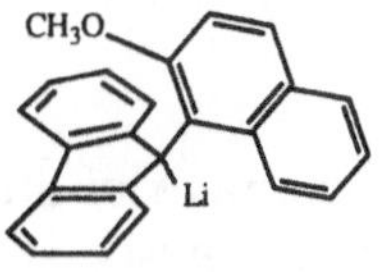   *ap*-**4-46**

The reason for the facile reactivity of the *sp* form should be deaggregation of butyllithium by the oxygen atom of the compound. The hexameric state of butyllithium in hydrocarbons changes to dimeric when it is dissolved in ether [29]. The oxygen atom in the solvent molecules ligates to lithium atom to deaggregate. In hydrocarbon solvents, this kind of deaggregation does not take place but it is expected that the ether oxygen of the methoxy group acts in such a way. Then butyllithium which is attached to the methoxy group in the *sp* form is close enough to deprotonate from the 9-position (*sp*-**4-47**), whereas a similar situation in the *ap* form is too far to deprotonate from the 9-position of the same molecule (*ap*-**4-47**).

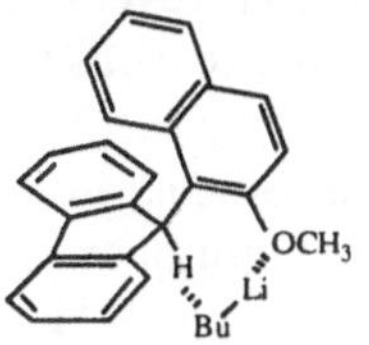

sp-**4-47** 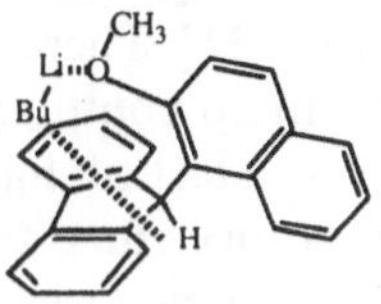   *ap*-**4-47**

A detailed study was carried out with these molecules and other related compounds. Firstly, 9-(2,6-xylyl)fluorene (**4-48**) was not deprotonated to a measurable extent even after several days in hexane-benzene at room temperature [26]. This is expected because the methyl group is larger than the π-system and thus blocks the approach of hexameric butyllithium to the 9-H efficiently. However, it reacted instantaneously if the reaction was carried out in tetrahydrofuran. This indicates that the deaggregation of butyllithium is a key factor for the reaction to occur.

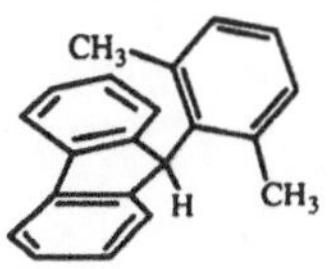

4-48

9-[(2-Substituted methyl)-6-methylphenyl]fluorenes (**4-49**), where the substituent is a dimethylamino, methoxy, or methylthio, were treated similarly. In every pair of the rotamers, the *sp* form was more reactive than the *ap* [30].

sp-**4-49**   *ap*-**4-49**

The detailed study indicated that the rates of lithiation are in conformity with the pseudo-first-order reaction in the *sp* isomer but are not so for the *ap* isomer. Study of the concentration effect on the reaction rates indicated that the reaction was second order in the *ap*-9-arylfluorenes [30]. Thus one is forced to consider that, in the case of *ap*, the deprotonation occurs in a molecule by assistance of another ligated molecule to butyllithium. Direct deprotonation by butyllithium is a slower reaction than the deprotonation with a butyllithium-ligated *ap*-9-arylfluorene.

The observed rate constants for the deprotonation are shown in Table 4.9. From these data, one can see a few points.

Firstly, the oxygen compound (**4-49**: X=O) is the most reactive among the compounds examined, both for the *sp* and *ap* series. It is reasonable to observe a larger rate constant for the oxygen compound than that for the sulfur compound (**4-49**: X=S), because the affinity of oxygen atom to lithium is higher than that of sulfur [31]. However, the order of amino-nitrogen and ether-oxygen is reverse of the known affinity toward the lithium ion [32]. There must be some reasons for the slower-than-expected rates for the amino compound.

The results are attributed to the steric effect on ligated forms. Since amino-nitrogens possess only one lone-pair of electrons, the dimethylamino group takes a conformation, in which the lone-pair electrons direct toward the 9-H of the fluorene ring to minimize the steric strain, in the free form. Ligating of this amine to anything larger than the lone-pair should destabilize the molecule. On the other hand, ether-oxygens possess two of the lone-pairs of electrons, one of which may be close to the 9-H and another of which is relatively far from the 9-H. Butyllithium can ligate to the latter lone-pair of electrons without much steric hindrance. Then the ligated form of the oxygen compound (**4-49**: X=O) must be more stable than the amino compound (**4-49**: X=CH$_3$N).

Table 4.9. Pseudo-first order rate constants for lithiation of compound **4-49** rotamers at 31.5 °C in benzene–hexane

X	k_{sp}/s^{-1}	k_{ap}/s^{-1}	k_{sp}/k_{ap}
NCH$_3$	1.25×10^{-2}	$< 5.3 \times 10^{-6}$	> 2300
O	3.80×10^{-2}	$< 1.6 \times 10^{-5}$	> 2300
S	1.95×10^{-4}	$< 8.0 \times 10^{-7}$	> 240

Secondly, from the second order rate constant, it is possible to estimate the maximum first order rate constant for deprotonation of the *ap* form. They are also listed in Table 4.9. These may be too large, if one judges from the fact that 9-(2,6-xylyl)fluorene (**4-48**) is not deprotonated to a measurable extent after several days, because both in the xylyl compound and in those in question here a methyl group or a substituted methyl group protects the 9-H. Yet, the relative rates, k_{sp}/k_{ap}, are estimated to be > 2300 for both the oxygen and the amino compounds and it is > 240 for the sulfur compound. The difference is indeed large.

In order to get information about the first-order rates for the *ap* compounds, kinetic runs at concentrations lower than before are necessary. It was not possible for the classical NMR spectrometer to do so but it became possible to do so by the use of modern NMR spectrometers and mass spectrometers.

This kind of investigation was carried out with 9-(2-substituted 1-naphthyl)-fluorenes (**4-50**) for the following reasons [33]. In the 9-(2,6-xylyl)fluorene and its derivatives, it was difficult to see the effect of the heteroatoms on the kinetics, because the parent compound did not show measurable reactivity. In contrast, though low, the reactivities of 9-(2-methyl-1-naphthyl)fluorene (**4-44**) are known and the rates of lithiation can be compared to discuss the heteroatom effect.

sp-**4-50**   ap-**4-50**

Due to the large difference in the reactivities of the *sp* and the *ap* rotamers, it was not possible to compare the rates at a given temperature. Thus for the *sp* isomer, the reactions were run at various temperatures and kinetic parameters were obtained. With the use of these parameters, the rates of lithiation at a given temperature are calculated. In actuality, however, there is another limitation. That is, the barrier to isomerization between the *sp* and the *ap* rotamers is not very high from the stand point of classical kinetics. It is especially so in the *ap* isomers, because their lithiation rates are low. As a compromise, the rates for the *ap* forms were measured at 35 °C only and those for the *sp* forms were obtained by calculation. The results are given in Table 4.10 [33], in which the data for the parent hydrocarbons are also presented. Here the second order means the first order each in the substrate and in the lithium compound. There are several points of interest in these data.

The rates of lithiation are not very much different when one goes from the hydrocarbon to those which carry heteroatoms, if one compares those of *ap* forms. Even for the *sp* series, the difference is only an order of 100, except for the methoxy compound, which gives ca. 10^4 times faster rates than those of the hydrocarbon.

Table 4.10. Second order rate constants ($\times 10^5/\text{s}^{-1}\text{mol}^{-1}\text{L}$) of lithiation of compounds **4-50** and **4-44** rotamers in benzene-hexane[a]

Form	4-50 (X=NCH$_3$)	4-50 (X=O)	4-50 (X=S)	4-44
sp	34.2	1090	13.1	0.14
ap	27.8	2.73	4.68	1.0

[a] The rates for compound **4-44** are those at 42 °C.

The difference in the *ap* series can be caused because of the difference in the size of the 2-substituent. If a butyllithium molecule ligates strongly, as in the case of the oxygen compound, the size should be larger than that which ligates weakly, such as in the sulfur compound. If the 2-substituent is bulky, the steric hindrance becomes large when the naphthyl group rotates to make a room for the attack by the butyllithium at the 9-position. Since a methyl group is very large, the parent hydrocarbon gives the smallest rates of lithiation.

The enhancement to a large extent in the case of *ap*-dimethylamino compound (**4-50**: X=NCH$_3$) is not well understood at the present time but it is possible that butyllithium ligated to the amino group of another molecule plays some role in the deprotonation of the compound.

The concentration effect on the rates of deprotonation was studied with the use of the *ap*-methoxy compound (**4-50**: X=O). At a level of 6 mmol L^{-1} concentration, which is the concentration used for other compounds as well, the *ap* methoxy compound gave a rate constant of 2.48×10^{-5} s^{-1}. The rate constant rose to 2.62×10^{-5} at 12 mmol L^{-1} and then 3.11×10^{-5} at 31 mmol L^{-1}.

The largest enhancement of the rates in the case of the *sp*-methoxy compound (**4-50**: X=O) may again be attributed to the presence of two pairs of electrons available for ligation, whereas there is only one pair in the amino compound, of which ligation to butyllithium is hampered by the steric effect.

The second order rate constant for the *ap*-methoxy compound (**4-50**: X=O) was obtained as 2.55×10^{-4} s^{-1} mol^{-1}L at 34 °C. This is a little lower than that for *ap*-9-(2-methoxymethyl-6-methylphenyl)fluorene (**4-49**: X=O). If the difference is significant, it will mean that the freedom of the side chain in the methoxymethyl compound (**4-49**) is responsible for this. Because of this freedom, the ligation in the ground state becomes preferable to the limited conformations in the methoxynaphthyl case as well as the transition energy for the abstraction. There may also be effects of the low basicity of the aromatic ether oxygen with respect to the aliphatic ether oxygen.

4.1.4 Other Types of Reaction

So far, reactions of rotational isomers of 9-arylfluorenes in which ionic species are involved have been discussed. There are other types of reactions which do not involve ionic species. In this section, reactions involving radicals and carbenes or nitrenes will be described.

4.1.4.1 Radical Reactions of 9-Arylfluorene Rotamers

Radicals are produced in various ways but, in studying the behavior of the rotameric radicals in 9-arylfluorenes, the maximum temperature that can be used must be taken into account. This is because, if rotation takes place, it is impossible to get reliable data. An example of this type of reaction is the decomposition of peroxyesters. tert-Butyl sp- and ap-1-(9-fluorenyl)-2-peroxy-naphthoates (4-51) were found to decompose at 80 °C with measurable rates. Rotational barriers for these compounds are known to be high enough to prevent rotational isomerization by heating at this temperature for several hours.

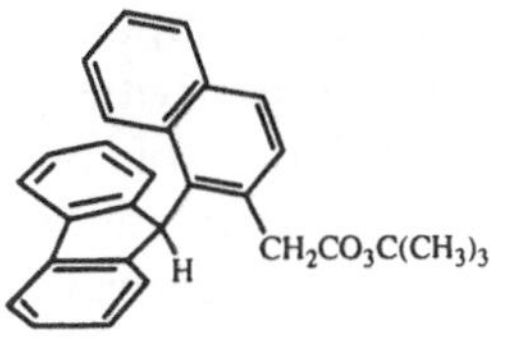

ap-4-51 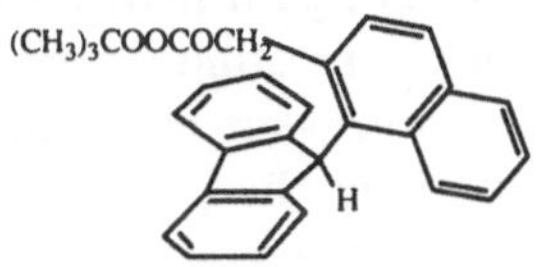   sp-4-51

Rates of decomposition of the isomeric peroxyesters in carbon tetrachloride were measured at 60.0–75.1 °C and at a concentration of 20 mmol L^{-1} [34]. At every temperature examined, the ap-peroxyester decomposes more rapidly than the sp by a factor of ca. 3. The products of decomposition were, as expected, compounds that were derived by the benzyl-type radicals which reacted with solvent molecules, 9-(2-chloromethyl-1-naphthyl)fluorene (4-52) and 9-[2-(2,2,2-trichloroethyl)-1-naphthyl]fluorene (4-53), and tert-butoxy radicals, 9-(2-tert-butoxymethyl-1-naphthyl)fluorene (4-54), though the formation of the trichloromethyl compounds (4-53) was at very low levels.

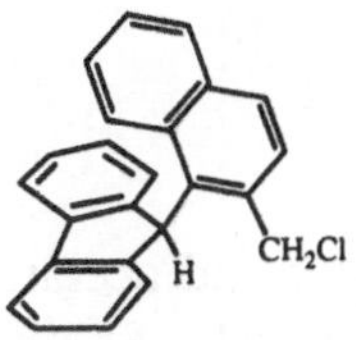

sp-4-52 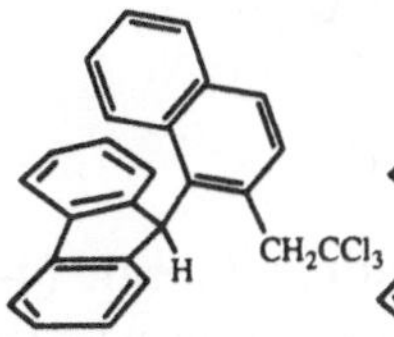   ap-4-53 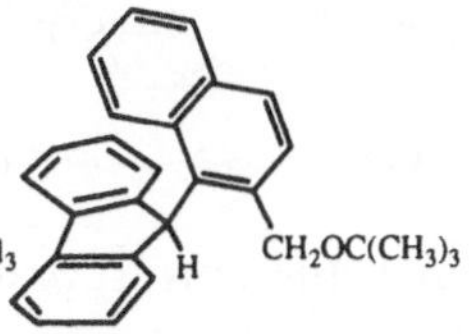   sp-4-54

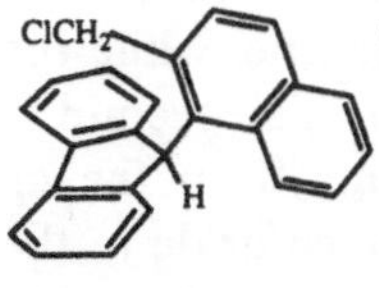

ap-4-52 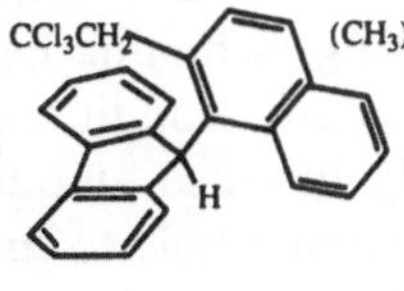   sp-4-53 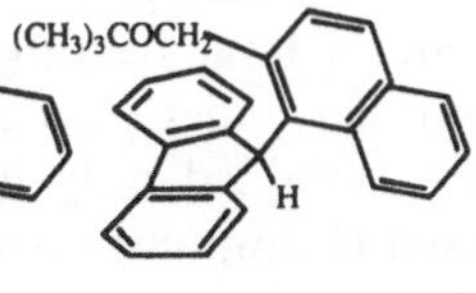   ap-4-54

The formation ratios of these compounds did not change to a large extent from one rotamer to another. However, it is significant that the yields of products derived from radicals and the solvent molecules are higher for the *ap*-peroxyester than for the *sp*, due to the steric effect of the fluorene moiety. No apparent effect of the π-system was observed.

The slower decomposition of the *sp*-peroxyester is rather surprising, because both steric congestion in the ground state and the possible π-participation should enhance the reaction rates in this conformation rather than the *ap*. One of the possibilities that might explain the slow decomposition of the *sp*-form would be the steric retardation of bond-lengthenings in the transition state of decomposition due to the presence of the fluorene ring. Solvent viscosity effects on the decomposition of compound **4-51** were studied and, with one exception, the decomposition was slow in a viscous solvent. It requires further study to understand the difference of the reactivity of the two peroxyesters.

Hunsdiecker reactions are another case that produces radicals, because its radical nature has been established, though it involves organolead compounds [35]. During such investigations, an interesting difference in the reactivities of rotamers was observed. The reaction was chlorodecarboxylation of 1-(9-fluorenyl)-2-naphthylacetic acid (**4-55**) in the presence of lead(IV) acetate and lithium chloride in benzene. This reaction is known to afford a chloro derivative in place of the carboxyl group as a main product and an acetate of the corresponding alcohol in a small amount.

ap-4-55   sp-4-52   sp-4-56

sp-4-55   ap-4-52   ap-4-56

The products were normal, the corresponding chloride (**4-52**) and acetate (**4-56**), when air was carefully purged and the reaction was run under argon, although the reactions showed differences in two points. One is that the reaction of the *sp* form is more sluggish than the *ap*. The former remained unreacted to the extent of ca. 35% even after 4 hour heating in boiling benzene, while the latter was completely decomposed under the same conditions. The second is that, whereas the acetate was obtained only in trace amounts from the *ap* carboxylic acid, it was obtained in amounts comparable to the chloride in the case of the *sp*. Since the acetate is known to be produced via carbocations, the results imply that the radicals formed over the fluorene ring have a longer life-time than that in proximity of the 9-H or the former is more easily oxidized than the latter. No clue to identify the cause is available as yet but it is tempting to attribute the cause to the π-participation if a radical is produced over the fluorene ring.

If oxygen is not carefully purged from the reaction system, the *ap* acid gives a peroxide, while the *sp* form is not affected to a significant extent. The structure of the peroxide was consistent with **4-61** and the mechanism of the formation are likely to involve an oxygen-trapped radical. Probably a 9-radical (**4-59**) formed by abstraction of hydrogen by the peroxy radical (**4-58**), which in turn is formed by capturing oxygen by the radical (**4-57**) that is produced by decomposition of the peroxyester, is stable enough to be oxidized by lead ions of high oxidation state to produce an intermediate cation (**4-60**) that cyclizes to the peroxy-oxygen.

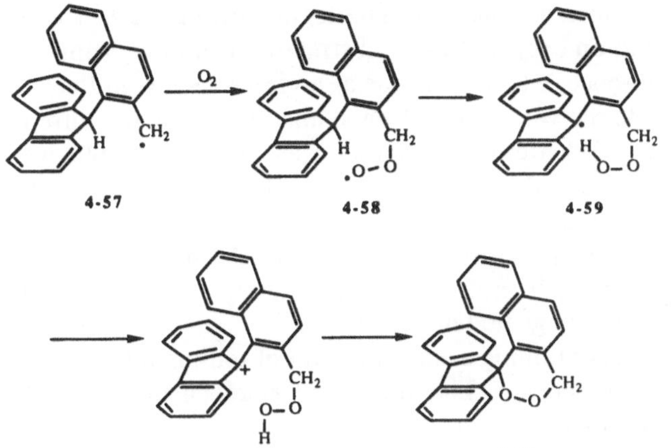

This hydrogen transfer to the peroxy-radical, which might also be formed over the fluorene ring, is not possible due to the long distance involved. Rather the benzylic radical or the organo-lead intermediate reacts with a chloro ligand. This is considered to be the cause for the difference for the two rotamers.

4.1.4.2 Nitrenes and Carbenes in Arylfluorene Rotamers

The photochemistry of 9-(2-azido-6-methylphenyl)fluorene (**4-62**) rotamers has been extensively studied [37]. Photolysis of the *ap* rotamer of the azide at cryogenic temperature showed the presence of the corresponding triplet nitrene (**4-63**) as well as an insertion product (**4-65**) which is easily produced due to the close proximity of the benzene ring of the fluorene. By contrast, the *sp* azide did not show the signal due to triplet nitrene (**4-63**) but absorption due to an

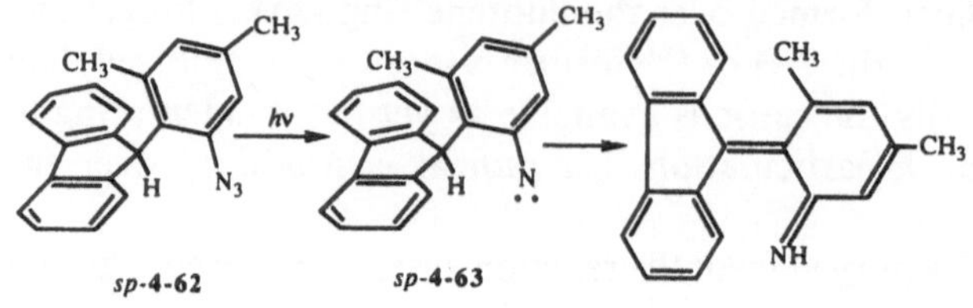

sp-4-62   sp-4-63   4-64

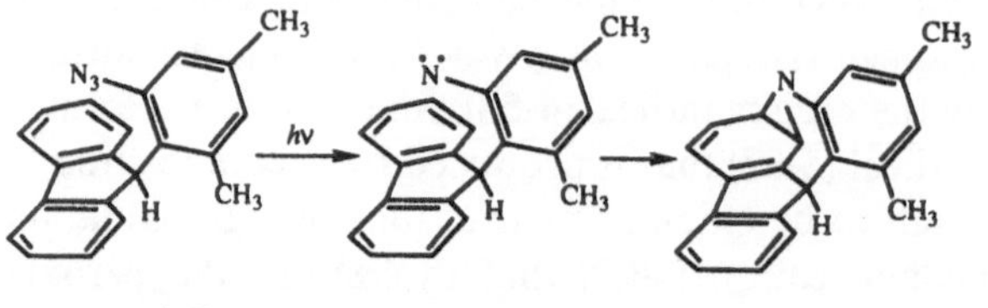

ap-4-62   ap-4-63   4-65

o-quinoneimine (**4-64**) was observed instead. The formation of the last compound must be attributed to an easy hydrogen abstraction from the 9-position of the fluorene ring. The absence of triplet nitrene as an intermediate in the case of *sp*-**4-62** photolysis reflects the very fast hydrogen abstraction by the nitrenic center and indeed, when a 9-deuterio compound was used, the triplet was observed. However, it was not possible to determine whether the easy hydrogen-abstraction was due to the tunnelling effect or not.

In fluid solutions at room temperature, those unstable intermediates react or rearrange to give stable compounds. But their reaction seems to be different to some extent from that at the cryogenic temperature.

The reaction in methanol containing a small amount of sodium methoxide is summarized in the following scheme. The rotational stereochemistry of the starting materials is not completely retained.

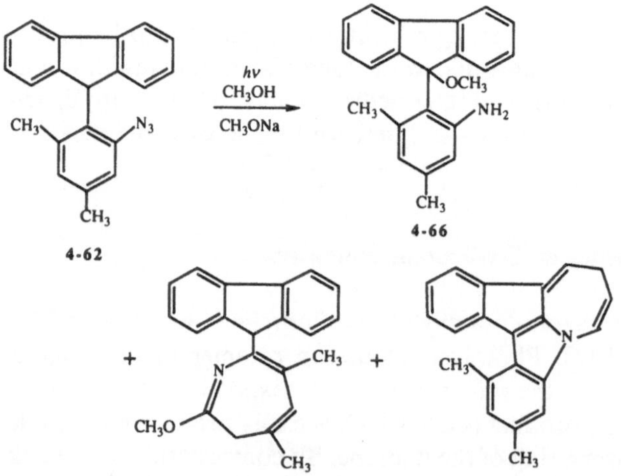

4-62   4-66

4-67   +   4-68

Formation of the methoxy-amine (4-66) is rationalized by considering the conjugate addition of methanol to the quinoneimine (4-64). The formation of the methoxyazepine (4-67) was assumed to proceed through an azanorcaradiene (4-69) of which formation is known in phenylnitrenes [38]. That of another azepine (4-68) is a result of base-catalyzed rearrangement of an azepine which is a valence isomer of (4-65).

It is understandable that the azepine (4-68) is formed as a major product from the *ap* isomer but it is difficult to understand why the azepine (4-68) is formed to an extent of almost 30% from the *sp* isomer. It is also difficult to explain the formation of the methoxy-amine (4-66) from *ap*-4-62 to an extent of ca. 10%. It is assumed, in order to explain the results, that the rotational barrier in the intermediate, an azanorcaradiene (4-69), is rather low. This is because the intramolecular addition in phenylnitrene is a reversible process and the stereochemistry can be scrambled to some extent by the following process.

Photolysis of the azide (4-62) in diethylamine affords products again derived by stereochemical scrambling to some extent. Interesting in this case, however, is that amines (4-70) are produced in considerable amounts. While the *ap*-azide gave *ap*-amine (4-70) and aryl-diethylamino compound (4-72) in 65 and 16% yields, respectively, the *sp*-azide afforded 11% *ap*-amine (4-70), 42% *sp*-amine (4-70), 10% 9-diethylamino compound (4-71) and 15% aryl-diethylamino compound (4-72). The primary amines are known to be produced from triplet nitrene [38] and indeed presence of oxygen or 1,3-pentadiene prohibited the formation of the primary amines. The formation of the 9-diethylamino compound (4-72) was rationalized on the basis of conjugate addition of diethylamine to the quinoneimine (4-64), and that of the *o*-diamine (4-72) by the addition of diethylamine to a benzazirine intermediate which is discussed below. It is interesting to note that *ap*-4-62 affords *ap*-4-70 only, whereas *sp*-4-62 gives both *ap*-4-70 and *sp*-4-70.

Most of the products are explained by assuming the same intermediates as discussed in the photolysis of the azides in methanol. However, the last compound (4-72) cannot be formed from the benzazirine (4-69) introduced

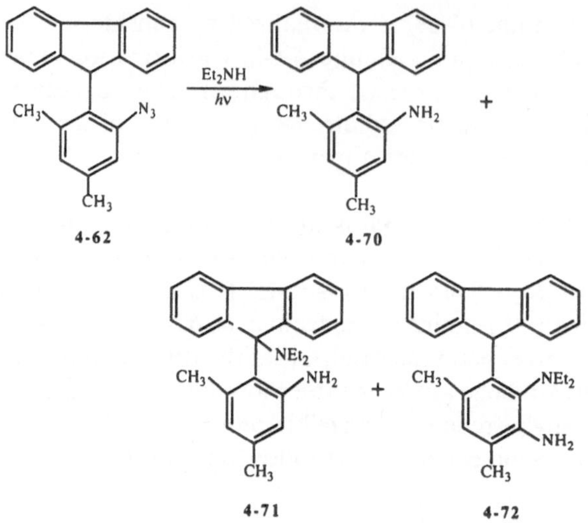

4-62   4-70

4-71   4-72

earlier. The intermediacy of another benzazirine (4-73) which is shown below should be assumed. It is postulated that the *sp*-nitrene in the singlet state reacts to produce the latter azirine, while its isomerization to the other azirine is possible. Diethylamine, being a very strong base, reacts with this former azirine, while methanol reacts with the latter azirine due to steric reasons.

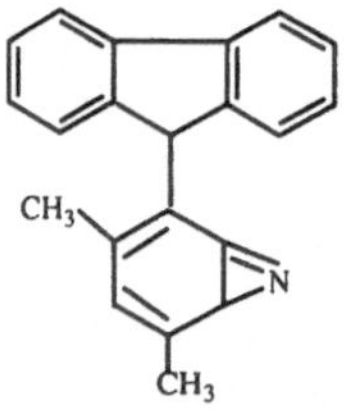

4-73

Rotationally isomeric carbenes show interesting differences as well in their reactivities in much the same way as nitrenes [39]. In hydrocarbons, both at low temperatures and at ambient temperatures, *sp* and *ap*-1-(9-fluorenyl)-2-naphthylcarbenes (4-75), that are produced by irradiation of the corresponding diazomethane (4-74), afforded single products, a norcaradiene (4-76) and a spiro compound (4-77), respectively. The formation of the norcaradiene is understood easily: addition of the singlet carbene to the benzene ring which is in proximity should produce 4-76. The formation of the spiro compound (4-77) can be attributed to the formation of quinodimethane by abstraction of the 9-H by the carbene.

In order to get an insight into this, an ESR study was carried out. In the *sp* diazomethane, the ESR spectra at 100–110 K showed the presence of a diradical after irradiation. Nonlinearity of the Curie plot suggests that the ground state is singlet. At higher temperatures, these signals were replaced by those due to a

ap-**4-74**   sp-**4-75**   **4-76**

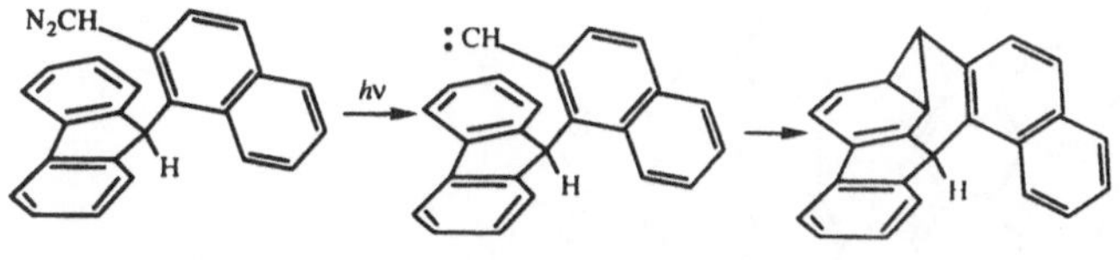

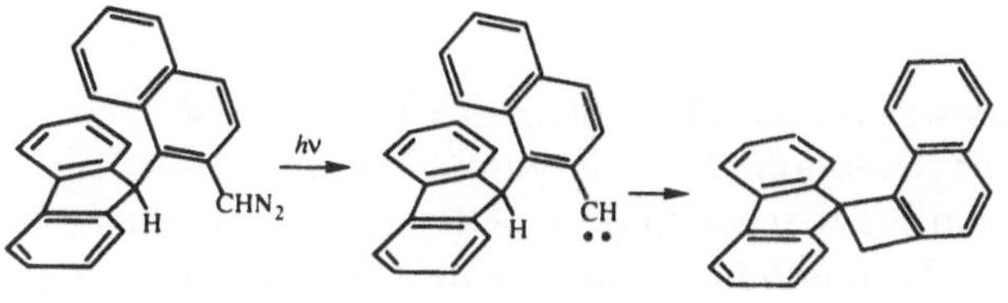

sp-**4-74**   ap-**4-75**   **4-77**

doublet species, which was assigned to a 9-fluorenyl radical produced by the abstraction of hydrogen by the carbenic center.

These results indicate that, when the carbene is produced from sp-diazomethane, it very easily abstracts the 9-H, which is close to it and quinodimethane (**4-78**) formed thus has the tendency to react in the following two structures. Formation of the spiro compound was attributed to cyclization of the biradical (**4-79**), although it is also possible that an electrocylic ring closure takes place.

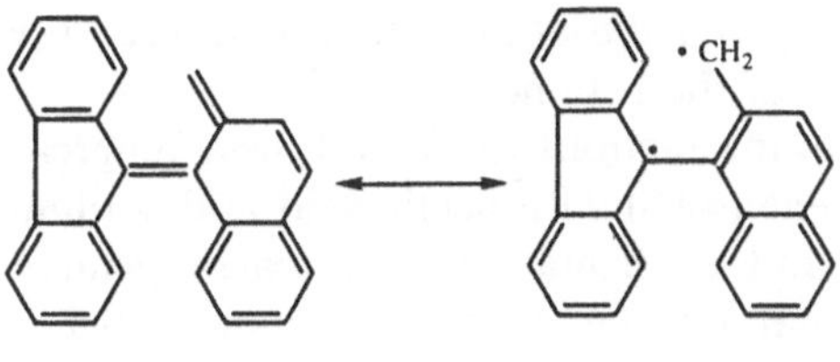

4-78   **4-79**

Photolysis of the diazo compound in ethanol was also carried out. From the ap-diazomethane, 9-(2-ethoxymethyl-1-naphthyl)fluorene (**4-80**) was obtained in addition to the norcaradiene (**4-76**) to indicate that the insertion of the carbene to the OH bond of ethanol can compete with the intramolecular addition.

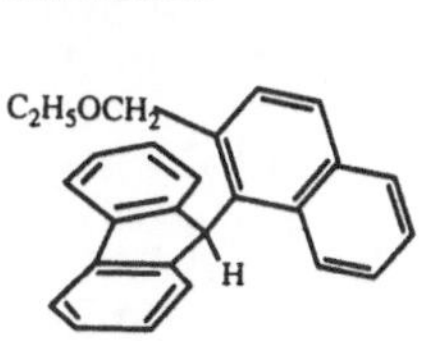

ap-**4-80**

Photolysis of sp-diazomethane in ethanol proceeded in much the same way, producing the benzocyclobutene derivative (**4-77**) and the ethoxymethyl compound (sp-**4-80**) when the reaction was carried out at 298 K. However, there was a difference when the reaction was carried out at 77 K. Among the products,

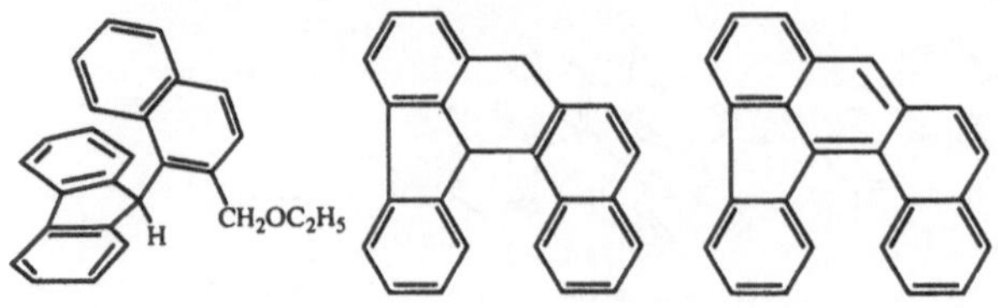

sp-4-80   4-81   4-82

another set of hydrocarbons was found. The dibenzoaceanthrylene (**4-82**) is an artifact: it is produced by air oxidation of the dihydro-compound (**4-81**).

The formation of dihydrodibenzoaceanthrylene (**4-81**) must be attributed to the electrocyclic reaction of the quinodimethane (**4-78**). However, the reason why this type of reaction occurs only in ethanol at a low temperature is not clear at the present time.

4.2 Reactivity of Substituted Triptycenes

As mentioned in Chap. 2, the distance between the 1-substituent and the 9-substituent in triptycenes is very short. It is ca. 2.7 Å between the α-atom of the 9-substituent and the 1-substituent, whereas it is ca. 3.0 Å between the β of the substituent which is in the sc-region of the 1-substituent and the 1-substituent. Therefore, the molecular interactions between the atoms mentioned are very strong and variously affect the reactivity of the rotamers.

A few of the typical examples of the interactions are as follows. Whereas diazotization of 9-(aminomethyl)triptycene (**4-83**) with acetic acid and sodium nitrite affords homotriptycyl alcohol and its acetate (**4-84**), rearrangement products, as sole products [40], a similar treatment of 9-(aminomethyl)-1,4-dimethoxytriptycene hydrochloride (**4-85**) gives unrearranged products (**4-88**) with a 32% yield together with a 37% yield of rearranged products (**4-86**) and a 6% yield of a cyclized product (**4-87**) [41].

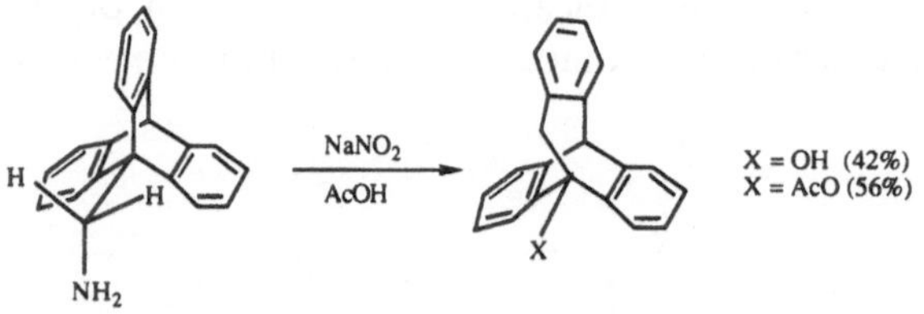

4-83   4-84

Although the detailed population ratios of rotamers of the reaction intermediates are not known for the dimethoxy compound, those which carry a positive charge, diazonium salt and the amine salt, are thought to exist as a mixture of rotamers in which the sc could predominate due to the molecular interactions, cation-methoxy or ammonium salt-methoxy. If a free carbocation

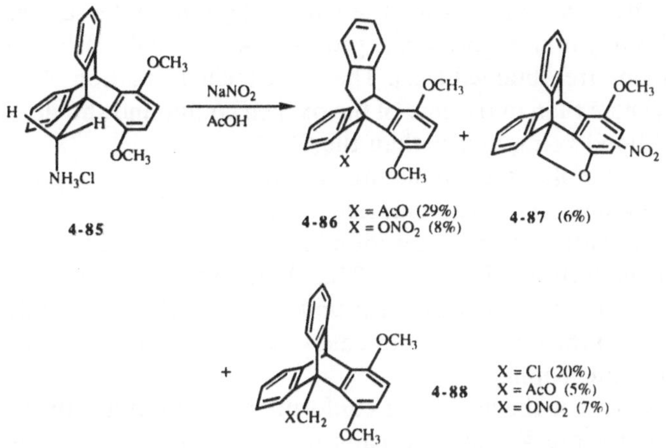

4-85

4-86 X = AcO (29%)
      X = ONO$_2$ (8%)

4-87 (6%)

4-88 X = Cl (20%)
      X = AcO (5%)
      X = ONO$_2$ (7%)

is produced from the diazonium salt, it confines to one conformation and is stabilized by the 1-methoxy group. Then the *ap*-direction is open to the attack of a nucleophile.

The *sc*-conformation should lead to the rearranged products because there is an anti-parallel C–C bond. It is interesting therefore to note that the non-rearranged products are produced in comparable amounts with the rearranged from the methoxy-substituted 9-aminomethyltriptycene, even though the *ap*-rotamer is not favored in conformational equilibrium in the intermediates. The results are indicative of a strong participation of the methoxy group to the reaction center. Formation of the cyclized product rather in a minor amount should be the result of the distance, 2.7 Å, which is too long for a C–O bond to be formed, although the strong intramolecular interaction between the cationic center and the 1-methoxy-oxygen is possible.

In the case of 9-(aminomethyl)-1,4-dimethyltriptycene hydrochloride (**4-89**), a cyclized product (**4-91**) is formed in almost 50% yield, whereas products (**4-90**)

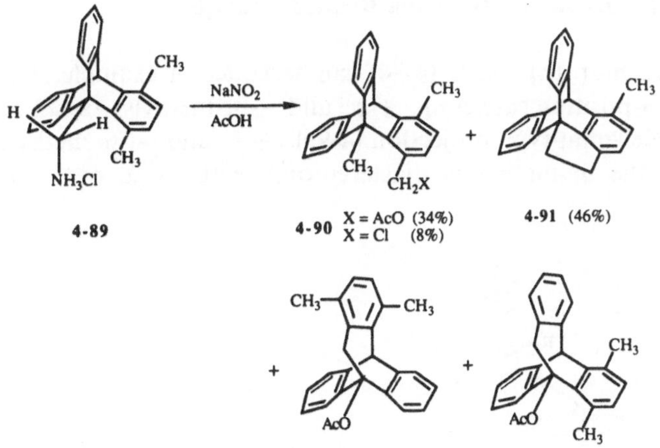

4-89

4-90 X = AcO (34%)
      X = Cl   (8%)

4-91 (46%)

4-92 (4%)

4-93 (2%)

which are derived after hydride transfer from the 1-methyl group to the cationic center amount to 42%. Only minor amounts of products (**4-92** and **4-93**) derived by skeletal rearrangement are obtained [42]. The more facile formation of the 5-membered ring (**4-91**) here than in the case of the oxygen compound should be attributed to the C–C bond being longer than the C–O.

In this case, no strong molecular interactions between the 1-methyl group and a diazonium species or the hydrochloride in the *sc* conformation is anticipated. Thus these substrates and the reaction intermediates should exist as a mixture in which the *ap* form predominates. The abundant formation of the products (**4-90**) derived by the hydride transfer from the 1-methyl group implies that the internal rotation in the reaction intermediates takes place more rapidly than the reaction with a nucleophile.

Since, in these cases, neither the molecular conformation is fixed nor the life times of the intermediates are known, we cannot proceed to discuss them further. It will clearly be understood that in the following examples the more detailed discussion is possible due to the fact that the conformation is fixed.

4.2.1 Radical-Forming Reactions

The three methyl groups in a *tert*-butyl are indistinguishable under normal reaction conditions and usually considered to be equivalent in reactivities. However, as is well known in the reactivities of three C–H bonds in a methyl in biological systems, the methyls in a *tert*-butyl should behave differently in those systems. This is because the methyl groups can be diastereotopic and/or enantiotopic, according to the steric environment in which the groups are located. Although, in chemical reactions under normal reaction conditions, it is not possible to distinguish the enantiotopic methyl groups, it should be possible to distinguish the diastereotopic pair of the methyls if internal rotation is frozen.

4.2.1.1 Substitution of a *tert*-Butyl Group and Related Reactions

9-*tert*-Butyl-1,2,3,4-tetrachlorotriptycene (**4-94**) can serve as an example. The *tert*-butyl group in the 9-position carries *ap*, + *sc*, and − *sc* methyls. As can be seen in the following Newman-type projection (**4-95**), + *sc* and − *sc* methyls are enantiotopic, but the *ap*-methyl is diastereotopic with + *sc* or − *sc*.

4-94   4-95

Therefore the reactivity of the *ap* methyl group must be different from those of the ± *sc*-methyls. Any reaction may serve as an example of this sort. Halogenation of methyls is one of them.

Chlorination of 9-*tert*-butyl-1,2,3,4-tetrachlorotriptycene (**4-94**) was carried out with sulfuryl chloride at 135 °C. This reaction proceeded to afford the methyl-chlorinated products without affecting the 10-CH group thanks to the pyramidal geometry, which is fixed due to the rigid triptycene skeleton and that does not contribute to the formation of stable radicals. The product ratio, the *sc*-chlorinated (*sc*-**4-96**) and the *ap*-chlorinated (*ap*-**4-96**), was 3.2. Since there are two *sc*-methyls and there is only one *ap* methyl, this result is interpreted as the *sc*-methyl is more reactive than the *ap* by a factor of 1.6 [43].

The higher reactivity of the *sc*-methyl can be attributed to the participation of the 1-chloro substituent, because it is the only difference for the two diastereotopic methyls that the *sc* has a chlorine atom in proximity.

The radical-forming transition state may be stabilized to some extent by the participation of the heteroatom in proximity. It is known that the *o*-substituent in *tert*-butyl peroxybenzoates facilitates the decomposition of the perester, when the heteroatom is in the third row of the periodic table [44]. The result mentioned here is consistent with the general view of the participation. However, there is one point which has to be cleared up before coming to this conclusion.

4-94   *ap*-4-96   *sc*-4-96

X-ray crystallography of 9-*tert*-butyl-1,2,3,4-tetrachlorotriptycene (**4-94**) revealed that the molecular structure of this was fairly distorted from the normal one, especially the bond angles. The *tert*-butyl group is tilted away from the 1-chloro substituent to relieve the steric strain [45]. This necessarily causes the *ap*-methyl group to be pushed into the triptycene skeleton. The structure can be illustrated in an exaggerated way by structure **4-97**. The *ap*-methyl is less open to the attack of radicals than the *sc*-methyls because of the steric environments. The low reactivity of the *ap*-methyl might be caused because of the steric protection.

Thus there are two possibilities that might give the observed results, enhanced reactivity of the *sc*-methyls with respect to the *ap*-methyl. They are the stabilization of the formed radicals due to 1-chloro participation (**4-98**) and the steric protection of the *ap*-methyl (**4-97**).

In order to diagnose which of the two, chlorine-participation or steric effects, is the main cause for the difference of the reactivities of the methyl groups,

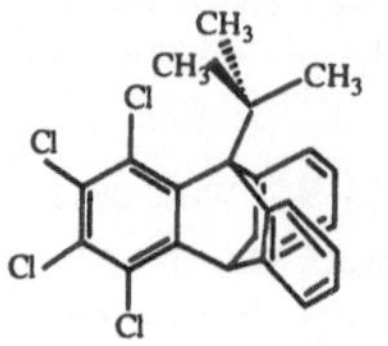

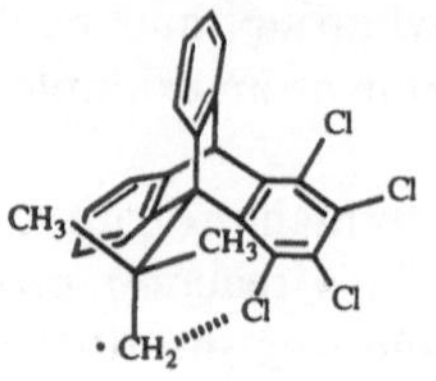

4-97   4-98

reactions of **4-94** with other halogenating reagents were performed and the results are shown in Table 4.11.

Chlorination by chlorine was found to be nondistinguishing. It gives a 2:1 mixture of *sc* and *ap* chlorinated compounds. The high reactivity of chlorine radical with respect to the radical produced from sulfuryl chloride must be responsible to the result. Indeed, less reactive bromine gives a higher *sc/ap* ratio 3.1, though the low reaction temperature, room temperature, with respect to the chlorination with sulfuryl chloride must be responsible to the results to some extent. Bromination of the compound with bromine-*N*-bromosuccinimide gave a relative rate of 1.5 for the *sc*-methyl.

A bromine substituent at the 1-position must activate the hydrogen abstraction more than a chlorine, if the heteroatom toward the bottom of the periodic table stabilizes the transition state more effectively than that toward the top. 1,2,3,4-Tetrabromo-9-*tert*-butyltriptycene (**4-99**) was brominated with bromine-*N*-bromosuccinimide to give ca. 10:1 reactivity of the *sc*-methyl relative to the *ap*. Thus practically complete chemoselectivity in the bromination is realized.

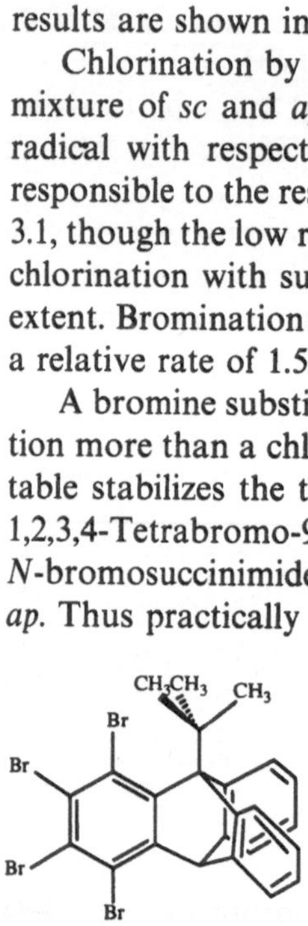

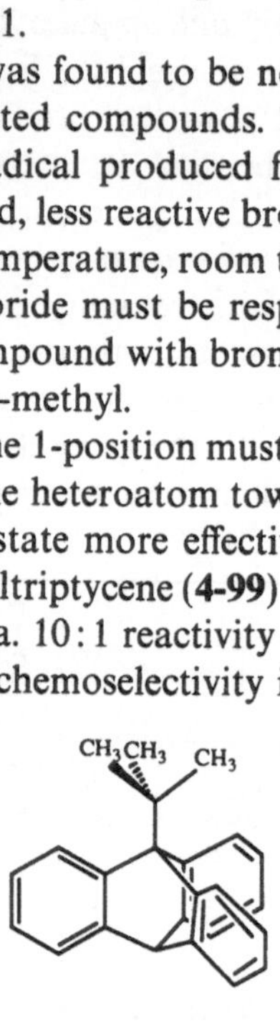

4-99   4-100

Table 4.11. Effects of halogenating reagents on the *sc/ap* ratios of 9-(2-halo-1,1-dimethylethyl)-1,2,3,4-tetrachlorotriptycene (**4-96**) in chlorobenzene or carbon tetrachloride

Reagent	*sc/ap*[a]
SO_2Cl_2	1.6 ± 0.1
Cl_2	1.0 ± 0.05
Br_2	3.1 ± 0.06
Br_2-NBS	1.5 ± 0.1

[a] Observed values are divided by two to accommodate the fact that there are two methyls in the *sc*-position

To the first approximation, the participation of the 1-substituent is responsible for the observed results. However, the detailed examination of the case by competitive reactions discloses a different aspect. The results shown in Table 4.12 indicate that, in chlorination with sulfuryl chloride, both sc and ap methyls are less reactive than the methyl in 9-*tert*-butyltriptycene (**4-100**). It is thought that the participation as well as the electronic effect is operative.

Since the hydrogen abstracting species here are electron-demanding, the retardation in chlorination of the *ap*-methyl must partly be attributed to the electron-withdrawing inductive effects of the four chlorine atoms in the benzene ring. Indeed, removing three chlorine atoms from the 2,3,4-positions raised the reactivity of the methyl groups to a considerable extent. Namely, 9-*tert*-butyl-1-chlorotriptycene (**4-101**) gave relative rates of chlorination of 0.94 and 0.92 for the *sc* and *ap* methyls to give *sc*-102 and *ap*-102, respectively, in chlorination with chlorine and 4.2 and 1.05 for the *sc* and *ap*, respectively in bromination with bromine.

4-101   ap-4-102   sc-4-102

The steric protection of the methyl seems to be a minor factor, though it is true that the *ap*-methyl is sterically protected with respect to the *sc*-methyl. This is because, if the steric factor were important, the reactivity of the *ap*-methyl would be different from one reagent to another due to the facts that the bond lengthening in the transition state, earliness or lateness, would be different and the different abstracting species would have different steric demand. The general tendency that bromination is more affected by the 1-substituent than the chlorination should be attributed to the low reactivity of the species concerned in hydrogen-abstraction.

Dichlorination of the *tert*-butyl group in compound **4-94** was also carried out. The first chlorination gives a mixture of *sc* and *ap* rotamers (**4-96**) and then

Table 4.12. Relative reactivities of methyls in 9-*tert*-butyltriptycenes relative to one of the methyls in 9-*tert*-butyltriptycene (**4-100**)

Halogenating Reagent	Compound	Relative Reactivities	
		sc	ap
SO$_2$Cl$_2$	**4-94**	0.95 ± 0.05	0.60 ± 0.03
Cl$_2$	**4-94**	0.84 ± 0.05	0.83 ± 0.05
Br$_2$	**4-94**	2.78 ± 0.3	0.89 ± 0.04
Br$_2$–NBS	**4-94**	1.16 ± 0.08	0.82 ± 0.04
Br$_2$–NBS	**4-99**	4.1 ± 0.8	0.8 ± 0.2

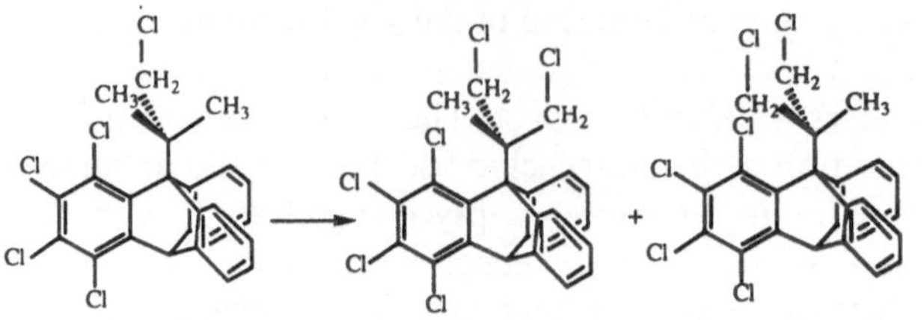

sc-4-96   sc-4-103   ap-4-103

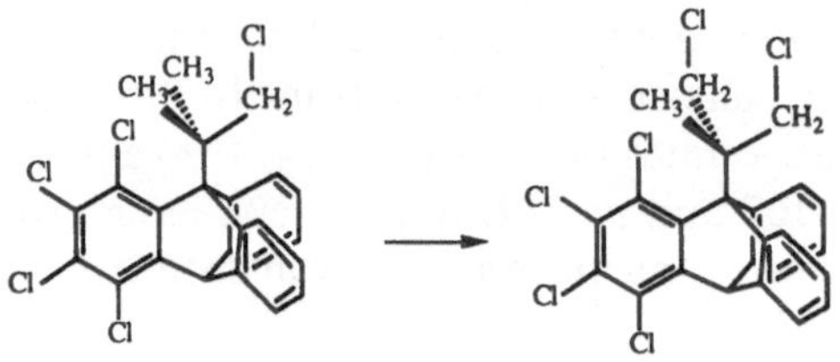

ap-4-96   sc-4-103

the second chlorination takes place on these isomers. The *ap*-monochloro compound (*ap*-**4-96**) possesses a pair of nondistinguishable *sc*-methyls, whereas the *sc*-monochloro compound (*sc*-**4-96**) has a diastereotopic pair, *sc* and *ap*, of methyls. It is expected that these diastereotopic methyls would behave differently in chlorination.

It was difficult to estimate the precise difference in the reactivities of these methyls due to the fact that the actual reaction was carried out as mixture of two monochloro compounds, but the formation ratio of the two dichloro compounds (*sc* and *ap*-**4-103**) was consistent with the assumption that the *sc*-methyl group has a reactivity of 1.6 times of that of the *ap* [46].

Dehalohydrogenation of organic halides with tin hydride is known to proceed with radical intermediates [47]. The rate-limiting step is abstraction of a halogen by a stannyl radical. This reduction of the rotameric chlorides obtained above was carried out with the use of tributyltin hydride and azobisisobutyronitrile in benzene [46]. Though this reaction was not very clean due to the fact that the aromatic chlorine substituent(s) is removed as well under the reaction conditions, reduction of a chloromethyl group in the *sc* position to a methyl was clearly faster than that in the *ap*.

Tributyltin hydride reduction of bis(chloromethyl) compounds (**4-103**) can be summarized in the following scheme. According to this scheme, one can analyze the results of competitive reactions and it was concluded that the *sc*-chloromethyl group is more reactive than the *ap* by a factor of ca. 7.

Reduction of mono(chloromethyl) compounds (**4-96**) gave a smaller difference in reactivities than the bis(chloromethyl) compounds. The factor was ca. 2. The diminished reactivity factor may involve the effects such as dipole interactions, which destabilizes the ground state of the bis(chloromethyl) compound, Cl–CH$_2$Cl interactions in the *sc*-chloromethyl group, which should be stronger in the bis(chloromethyl) compound than in the mono(chloromethyl), and increased electron demand at the reaction center derived from the bis(chloromethyl). Detailed discussion is not possible at present due to the presence of unknown factors.

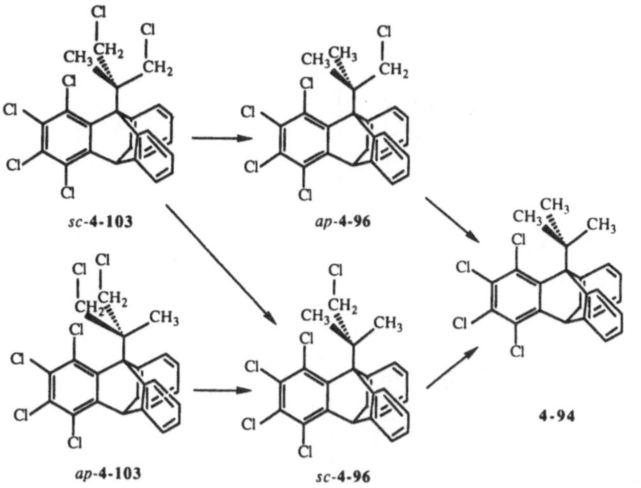

4.2.1.2 Thermolysis of Peroxyesters

Decomposition of peroxyesters is another route to give radicals. These peroxyesters should be prepared if carboxylic acids become available. These carboxylic acids (4-108) were prepared via rather a long route that was necessary for the separation of the rotamers. The olefinic triptycene (4-104), that was prepared from the corresponding anthracene and benzyne, was dihydroxylated with osmium tetroxide in the presence of amine oxide and the diols (4-105) were converted to acetonides (4-106). Chromatography of the acetonides was able to separate the rotamers about the C_9-C_{subst} bond, though the sc form consisted of two diastereomers due to the fact that the acetonide contained a chiral center in addition to the chiral rotational axis. Hydrolysis of the acetonides followed by oxidation of the diol (4-107) with periodic acid and then by oxidation of the resulting aldehyde with potassium permanganate afforded the desired carboxylic acid (4-108). In the case of the tetrahalo compounds where the halogens are chlorine or fluorine, it is possible to separate rotamers of olefins by chromatography on silver nitrate-impregnated silica gel.

In principle, these reactions leading to the carboxylic acid (4-108) are also examples of those of diastereomeric rotamers and thus should show differences from one rotamer to another. However, there was no appreciable difference noted except that the oxidations of sc-forms of the diol and the aldehyde can be a little slower than the ap-isomer due to the steric effect.

Decomposition of peroxy esters is known to proceed via radical mechanisms. Although it is now known that the scission of the O–O bond in the *ap*-4-104 *sc*-4-104

OsO_4, $(CH_3)_3NO$

ap-4-105 *sc*-4-105

$(CH_3)_2C(OCH_3)_2$
TsOH

ap-4-106 *sc*-4-106

peroxyester is the initiation step which is closely followed by a C–C bond breakage to eject carbon dioxide [48], the effects of substituents, which are close to the α-carbon to the carbonyl of the peroxy ester but not to the peroxy oxygen, are known in decomposition of *tert*-butyl *o*-substituted peroxybenzoates [44]. The 1-substituent effect on the decomposition of the peroxyester rotamers is expected on this basis.

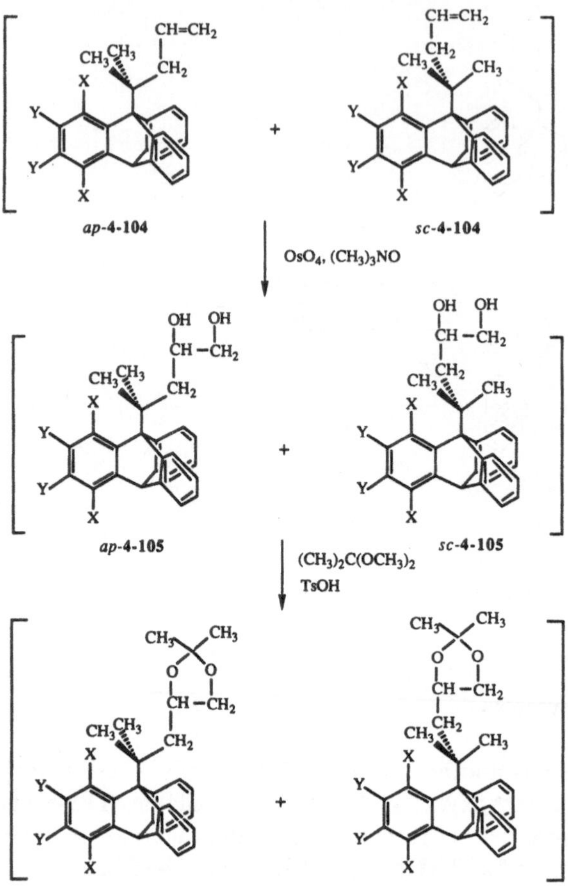

tert-Butyl 3-(substituted 9-triptycyl)-3-methylperoxybutyrates (**4-109**) were decomposed by heating in toluene solutions. The products of the thermolyses were, as expected, 9-*tert*-butyltriptycenes (**4-112**) derived by the hydrogen abstraction by the formed radicals (**4-110**), 9-(2-*tert*-butoxy-1,1-dimethylethyl)-triptycenes (**4-111**), a combination product from the *tert*-butoxyl and **4-110** radicals, and cyclization products (**4-113**) that are produced by the attack of the intermediate radicals (**4-110**) on one of the *peri*-positions of the benzeno bridges, in addition to the parent carboxylic acids that are apparently produced by the induced decomposition, when the substituent was fluoro, chloro, bromo, meth-

ap-4-107

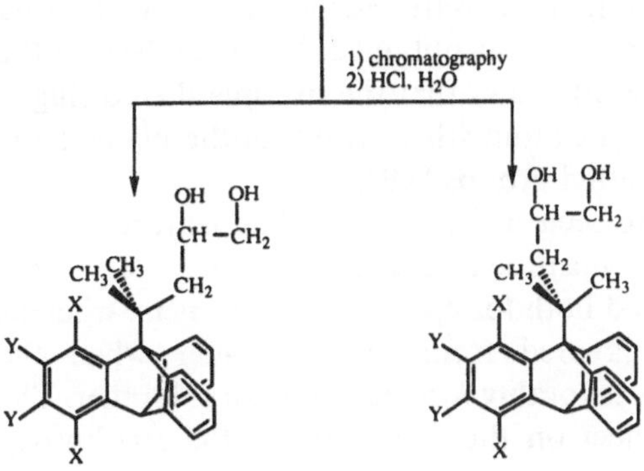

sc-4-107

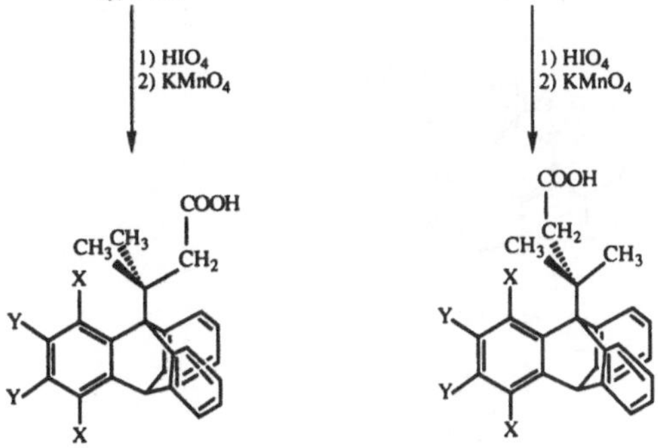

ap-4-108

sc-4-108

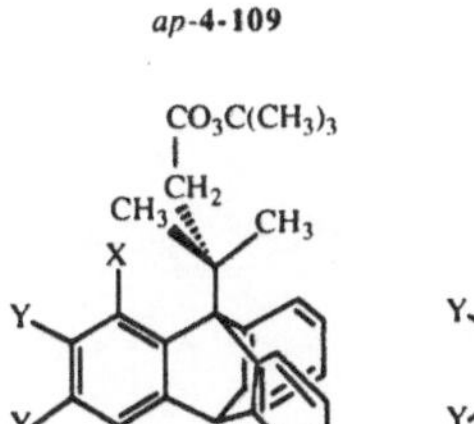

ap-4-109

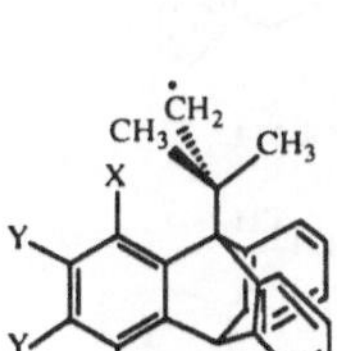

ap-4-110

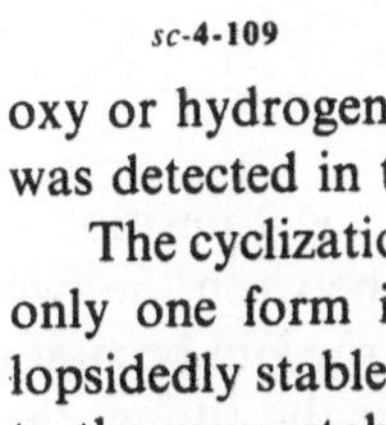

sc-4-109

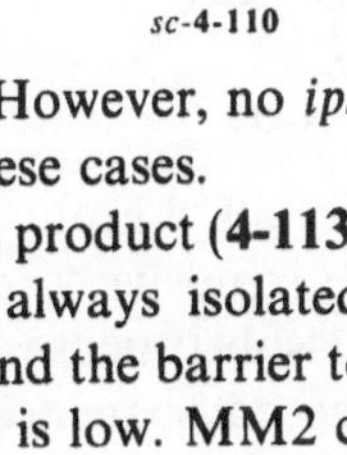

sc-4-110

oxy or hydrogen. However, no *ipso*-attack of the intermediate radicals (**4-110**) was detected in these cases.

The cyclization product (**4-113**) is expected to exist as rotational isomers but only one form is always isolated, probably because one of the rotamers is lopsidedly stable and the barrier to the transformation from an unstable isomer to the more stable is low. MM2 calculation shows that the stable form is that

derived from the *ap* isomer. These compounds were independently synthesized
and it was shown that only one compound was obtained. X-ray analysis of the
tetrachloro compound (**4-113**: X=Y=Cl) shows that the five-membered ring is
puckered and forms an envelope but the atom which is out of the plane is the
quaternary carbon that carry two methyl groups [49].

The product formation ratios are seen in Table 4.13. The features of the
products are as follows. When the 1-substituent is small, the formation of the
tert-butyl compound (**4-112**) is favored both for *ap* and *sc* conformers, whereas
the 5-membered cycle (**4-113**) is favored from the *ap*-isomer when the
1-substituent is large, though it is less favorably formed from the *sc*-isomer. The
1-substituent seems to have little effect on the formation of the *tert*-butoxy
compounds (**4-111**) [50].

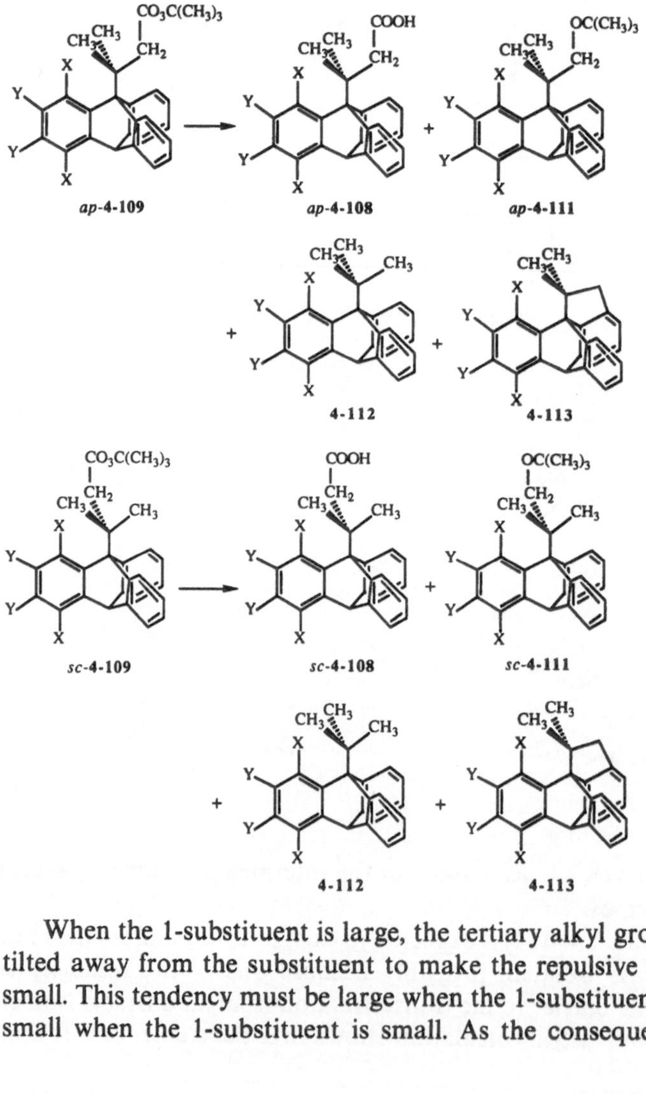

When the 1-substituent is large, the tertiary alkyl group at the 9-position is
tilted away from the substituent to make the repulsive force between the two
small. This tendency must be large when the 1-substituent is large and becomes
small when the 1-substituent is small. As the consequence of the tilting, the

Table 4.13. Product distribution in the thermolysis of *tert*-butyl peroxyesters (**4-109**) in toluene at 120 °C (% yield)

X	Y	Form	4-108	4-111	4-112	4-113
H	H	—	8	16	49	0
F	F	*ap*	3	17	40	14
		sc	19	10	49	0
Cl	Cl	*ap*	17	24	trace	48
		sc	12	14	16	32
CH₃O	H	*ap*	trace	23	31	36
		sc	3	19	43	19
CH₃	H	*ap*	trace	28	5	55
		sc[a)]	trace	11	3	0

[a)] Main products are those derived by rearrangement. See text.

radical center formed by the decomposition is pushed into the triptycene skeleton in the case of *ap*. This should be favorable for the formation of the five-membered ring (**4-113**) because the benzeno bridge that reacts with the radical center is closer than the case in which the *tert*-alkyl group is not tilted. The yields of the 5-membered ring are consistent with this consideration. On the other hand, if the radical center is exposed to the outer environments, this should favor the formation of hydrogen-abstraction and combination with the *tert*-butoxy radical. Reduced formation of the 5-membered rings from the *sc*-radicals is further support for the interpretation.

When one compares the product ratios from the *ap* and the *sc* isomers (**4-109**), one notices that the formation of the 5-membered cycle (**4-113**) is always less in the *sc* than in the *ap*. The same trend is seen also in the formation of the *tert*-butoxy compound (**4-111**). As compensation, the *tert*-butyl compounds (**4-112**) are formed more abundantly in the *sc* than in the *ap*. It is tempting to consider that these differences are derived from the substituent effect but decisive support is lacking for this conclusion.

The rates of decomposition of these peroxyesters are compared in Table 4.14. Generally the rotamers in a compound behave very similarly, though *ap* isomers show a tendency of decomposing faster than the *sc*. Especially interesting is the

Table 4.14. Rates of decomposition of *tert*-butyl peroxyesters (**4-109**) in toluene at 120 °C ($\times 10^4 \, s^{-1}$)

X	Y	*ap*	*·sc*
H	H	2.17	2.17
F	F	1.91	1.56
Cl	Cl	3.58	2.77
CH₃O	H	3.04	3.90
CH₃	H	2.42	2.27

fact that the series of compounds in the 1-halo derivatives shows little effect of the substituent. This is in sharp contrast with the *o*-substituted peroxybenzoates. Probably other factors, such as steric, are more important in determining the rates than the halogen participation.

The *sc*-1,4-dimethyl peroxyester (**4-114**) behaves completely differently from the others, although the *ap* isomer gave similar results to the others. The *ap*-isomer afforded, on thermolysis, a trace of the carboxylic acid (*ap*-**4-108**), X=CH$_3$, Y=H), 28% of *tert*-butoxy compound (*ap*-**4-111**, X=CH$_3$, Y=H), 5% of the *tert*-butyltriptycene (**4-112**, X=CH$_3$, Y=H), and 55% of the cyclized compound (**4-113**, X=CH$_3$, Y=H). By contrast, the *sc*-isomer (**4-114**) gave a trace of the carboxylic acid (*sc*-**4-108**, X=CH$_3$, Y=H), 11% of the *tert*-butoxy compound (*sc*-**4-111**, X=CH$_3$, Y=H), and 3% of the *tert*-butyltriptycene (**4-112**, X=CH$_3$, Y=H), as the products similar to others. No cyclic compound (**4-113**, X=CH$_3$, Y=H) was detected. Other products were identified as 7% 1-*tert*-butoxymethyl-9-*tert*-butyl-4-methyltriptycene (**4-115**), 40% 9-*tert*-butyl-4-methyl-1-(2-phenylethyl)triptycene (**4-116**), and 29% 1,2-bis(9-*tert*-butyl-4-methyl-1-triptycyl)ethane(**4-117**).

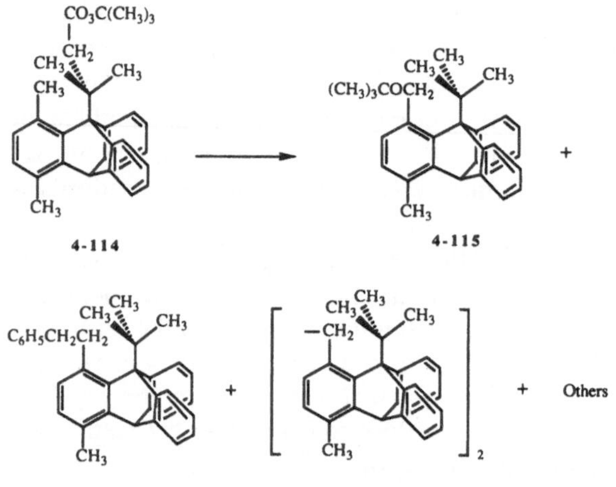

4-114   4-115

4-116   4-117   + Others

The formation of the last-mentioned compounds (**4-115** to **4-117**) can only be rationalized by assuming that the hydrogen transfer from the 1-methyl group to the radical center formed by decomposition of the peroxyester moiety is facile. Namely the hydrogen transfer causes the isomerization of **4-118** to **4-119**. Then

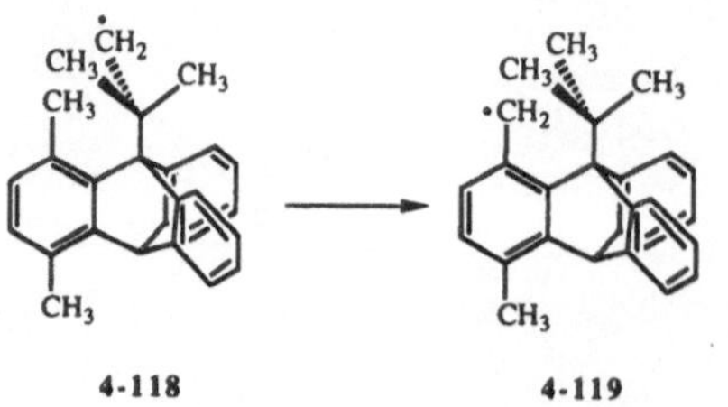

4-118   4-119

the 1-triptycylmethyl radical (**4-119**) reacts with the *tert*-butoxy radical also formed by the thermolysis to form compound **4-115**, with the benzyl radical formed by abstraction of hydrogen from the solvent to afford compound **4-116**, and dimerizes to form compound **4-117**.

This facile hydrogen transfer is reasonable because the benzylic radical must be more stable than a primary alkyl radical and the 1-methyl group and the radical center in the *sc* region must be very close to each other. However, formation of the *tert*-butoxy compound (*sc*-**4-111**) in a considerable amount and probably that of the *tert*-butyltriptycene (**4-112**), though in a reduced amount, speak for the hydrogen transfer to be energy-requiring. The rates of the hydrogen transfer must be comparable to the life-time of the *tert*-butoxy radical, because the 1-*tert*-butoxymethyl compound (**4-115**) was formed in a fair amount.

4.2.2 Cation-Forming Reactions

4.2.2.1 Lewis Acid-Catalyzed Ionization of Halogen Compounds

The first sign of a difference in cation-forming reactions of rotamers of triptycene derivatives was observed when separation of the rotamers of 1,2,3,4-tetrachloro-9-(2-chloro-1,1-dimethylethyl)triptycene (**4-96**) was attempted by chromatography on silica gel. The *ap*-isomer was separated without difficulty but the *sc*-isomer decomposed under the conditions [51]. The results imply that the cation formation from the *sc*-isomer was facilitated to give decomposed products during the chromatography.

Mixtures of *ap* and *sc* rotamers of compound **4-96** were treated with Lewis acids. Whereas the *sc* form reacted with a pseudo-first order rate constant of 5.3×10^{-4} s^{-1} at 41 °C when titanium(IV) chloride was used in excess, the *ap*-form did not show any measurable rate, being less than 8.7×10^{-8} s^{-1}. When a stronger Lewis acid, antimony(V) chloride, was used, the *sc*-form reacted with a pseudo-first order rate constant of 8.8×10^{-4} s^{-1} at 39.8 °C. Under the same conditions the *sc*-form reacted almost instantaneously: the rate constant was too large to measure.

The products were the same mixture of cyclized compounds (**4-123**), the structures of which were established by independent syntheses, after Wagner-Meerwein rearrangements.

It is assumed that a carbocation (**4-120**) is formed by assistance of the Lewis acid. The cation apparently undergoes two consecutive Wagner-Meerwein rearrangement, when antimony(V) chloride is used. The cations (**4-121**) produced by the rearrangement do not explain the experimental results and further rearrangement to produce cations (**4-122**) is necessary. Then the reaction scheme can be written as shown.

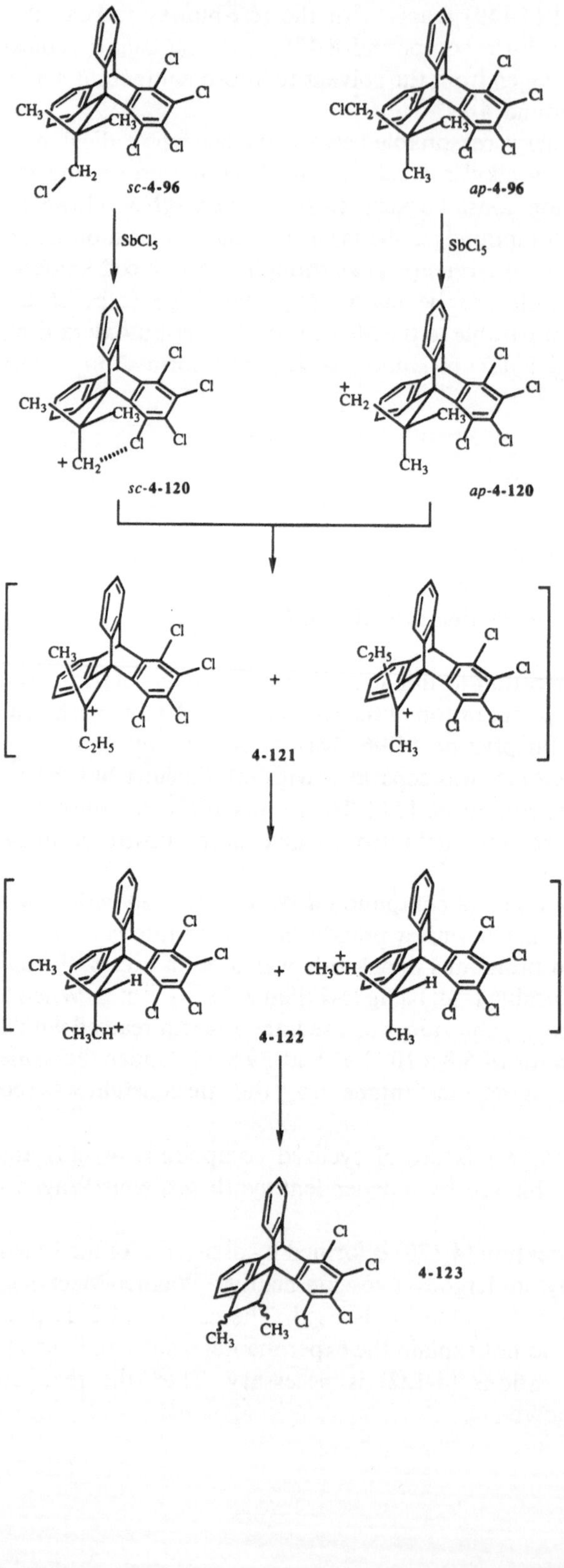

sc-4-96   ap-4-96

SbCl₅   SbCl₅

sc-4-120   ap-4-120

4-121

4-122

4-123

It is evident that the intervening cation can be stabilized by the participation of the chloro substituent which is closely located at the cationic center in the case of *sc* (*sc*-**4-120**) but such assistance is not possible for the *ap* cation (*ap*-**4-120**). Thus the ionization of the *sc* chloride is facile, whereas the *ap* is not.

After the ionization, the cation loses its stereochemistry because of the multiple pathways in the rearrangement and because internal rotation for compounds having an sp^2-sp^3 bond is fast at room temperature even for the triptycene derivatives. Thus the cyclized product (**4-123**) is a mixture of dimethyl compounds in which one can see every stereochemistry which is theoretically possible.

If a less active Lewis acid is used in the ionization reactions, the *sc* form produces olefins which can be derived from the intervening cations, whereas the *ap* remains intact. For example, a reaction of a mixture of the rotameric chloride (**4-96**) with silver nitrate gives *E* and *Z*-9-(1-methyl-1-propenyl)-1,2,3,4-tetrachlorotriptycene (**4-124** and **4-125**) and unreacted *ap*-**4-96**. Titanium(IV) chloride is an interim Lewis acid in a sense that it gives both the olefins and the cyclized product, when it was used in a small excess.

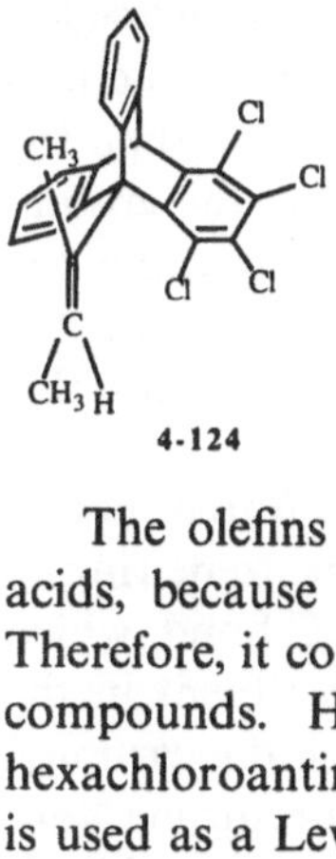

4-124 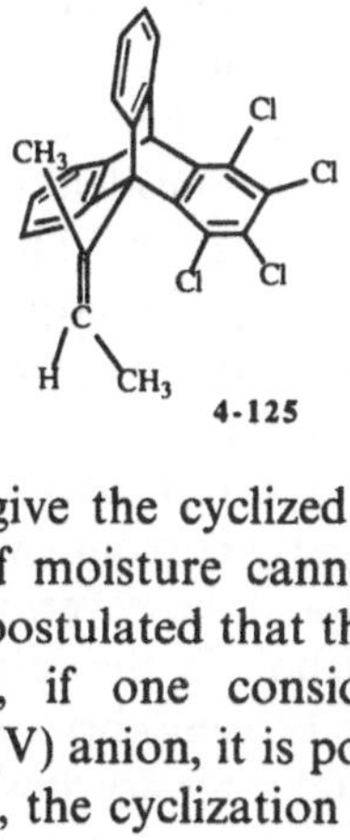   4-125

The olefins could give the cyclized products if treated with strong Lewis acids, because trace of moisture cannot be removed from organic solvents. Therefore, it could be postulated that the olefins are intermediates to the cyclic compounds. However, if one considers the smallest proton affinity of hexachloroantimonate(V) anion, it is possible that, when antimony(V) chloride is used as a Lewis acid, the cyclization takes place without intervention of the olefins.

A further example of ionization of a halide with a silver salt is described in Sect. 4.2.2.3.

4.2.2.2 Intramolecular Friedel-Crafts Acylation

Another typical cation-forming reaction is a Friedel-Crafts acylation reaction. This reaction was carried out with the rotameric carboxylic acids. The acids (**4-108**) were treated with oxalyl dichloride and the formed acid chloride was subjected to the action of titanium (IV) chloride.

The tetrahalo carboxylic acids (**4-108**, X=F or Cl) underwent normal Friedel-Crafts cyclization and afforded rotameric cyclic ketones (**4-127**) [52], as can be seen in the following scheme where Newman-type projections are used to

ap-4-126   ap-4-127

sc-4-126   sc-4-127

illustrate the difference and relations in stereochemistry both in carboxylic acids (4-126) and in the ketones (4-127).

Interestingly, these ketones show rather low barriers to isomerization and isomerize at around 100 °C with measurable rates. The low barrier to isomerization is attributed to the high energy ground state due to the structural requirement. The fluoro compound gives a higher barrier to rotation than the chloro compound. This is reasonable if one considers relaxation of the ground state by replacing a small substituent for a large one. The population ratio of the ketones also shows an interesting tendency. The larger the 1-substituent, the larger the population ratio, the sc form being preferred. The reason for this preference for the sc form is not known [52].

The structures of these ketones are interesting, because their infrared spectra indicate that the carbonyl groups are nearly coplanar with the benzene ring to which they are attached. On the other hand, however, if the C–C bond, which connects the carbonyl group to the carbon in the tertiary substituent in the original compound, is eclipsing with the benzeno bridge to which the carbonyl group is attached, the molecule must take a very unstable all-eclipsing conformation. Molecular mechanics calculation indicates that a set of three bonds for the substituent-derived part and another set of three which are a part of the triptycene skeleton form a nearly staggered conformation. If this calculation is correct, then the benzene ring which bears the carbonyl group must be distorted considerably. X-ray crystallographic investigations of these molecules are awaited.

The intramolecular Friedel-Crafts acylation of 3-(1,4-dimethoxy-9-triptycyl)-3-methylbutanoic acid (4-128) showed another interesting aspect of the rotational isomer chemistry. When the ap-acid was treated as in the case of others, it gave the corresponding ketone (4-129). By contrast, the sc-acid afforded a lactone (4-130), in addition to the corresponding ketone [53].

The formation of the lactone must be attributed to the attack of the carbonyl cation on the 1-methoxy-oxygen followed by removal of the methyl group from the methoxonium cation by a nucleophile. The ketones derived from the isomeric carboxylic acids are isomeric as well in this case and the barriers to isomerization are 29.9 and 31.2 kcal/mol for the $ap \rightarrow sc$ and $sc \rightarrow ap$ processes, respectively.

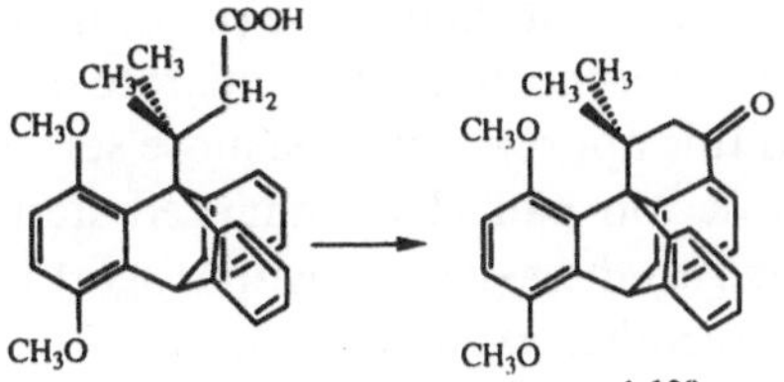

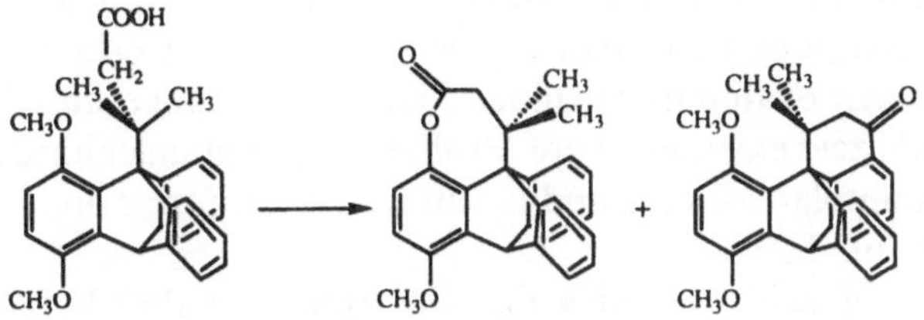

The formation of the lactone and the ketone is competitive. The formation ratio is 42 in favor of the ketone when the reaction of sc-**4-128** in dichloromethane was carried out at room temperature in the presence of 2 mol of titanium(IV) chloride. The ketone is an almost exclusive product of this reaction to indicate that the basicity of the π-system of the triptycene skeleton and the steric requirements are favorable for ketone-formation. When the reaction was carried out under other conditions, it provides different ratios of the lactone to the ketone as shown in Table 4.15.

The general tendency is that, as the solvent polarity is decreased, the formation of the lactone is enhanced, and, as the Lewis acid acidity is increased, the formation of the lactone is favored. The results may be interpreted in the following way.

The acid chloride and the Lewis acid produce a tight ion pair in these solvents. However, the tightness of the ion pair is different from one solvent to

Table 4.15. Effects of solvent and catalyst on the formation ratios of ketone sc-**4-129** to lactone **4-130**

Solvent	Lewis Acid	sc-**4-129**/**4-130**
CH_2Cl_2	$TiCl_4$	42
C_6H_6	$TiCl_4$	6.0
C_6H_{12} [a)]	$TiCl_4$	1.3
CH_2Cl_2	$SnCl_4$	1.7
C_6H_6	$SnCl_4$	0.42
C_6H_{12} [a)]	$SnCl_4$	0.18
C_6H_6	$AlCl_3$	0.84
C_6H_{12} [a)]	$AlCl_3$	0.49

[a)] cyclohexane

another. Ion pairs must be tighter in a nonpolar solvent than in the polar. If the ion pair is to react with the π-base to afford the ketone, it has to penetrate into the triptycene skeleton and the steric effect to this approach will be more severe for a very tight ion pair than that for a loose ion pair. The enhanced steric demand for the approach of the tighter ion pair to the π-system, with respect to the approach to the methoxy-oxygen, which is rather exposed, is the cause for the favored lactone formation in nonpolar solvents. The same explanation may apply for the favored formation of the lactone when a weak Lewis acid is used. The stronger Lewis acid gives looser ion pairs, thus reducing the steric hindrance for the reaction. The case of aluminum chloride is an exception because it is usually the strongest Lewis acid examined here. Probably, aluminum chloride is a dimer in these rather nonpolar solvents and is still so when it forms an ion pair after reacting with organic halides.

Friedel-Crafts reactions of acid chlorides derived from 3-methyl-3-(1,4-dimethyl-9-triptycyl)butanoic acid (**4-131**) show another interesting point of difference in the reactivity of rotamers [53]. The *ap*-isomer gives the expected *ap*-ketone (**4-132**) because the two benzeno bridges which can react with the acyl cation derived from the chloride are enantiotopic: the ketone is identical if we neglect the presence of enantiomers.

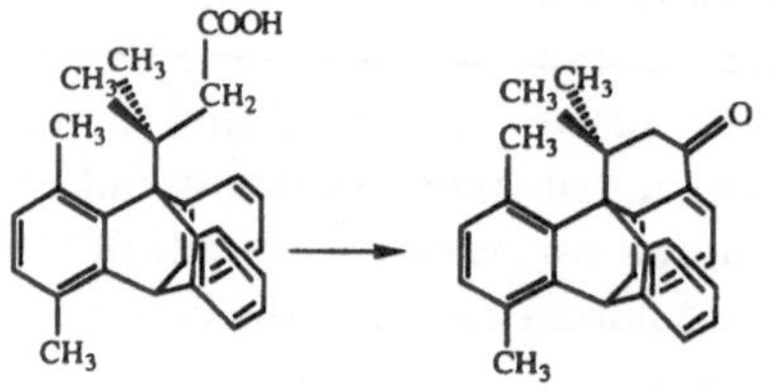

ap-**4-131**   ap-**4-132**

By contrast, the *sc*-acid chloride has two benzeno bridges to react which are different. Therefore, formation of two products was expected, if an ipso attack was possible on the 1,4-dimethylbenzeno bridge. The product from the *sc*-carboxylic acid (**4-131**) was indeed a mixture of two ketones, one of which was the rotamer (*sc*-**4-132**) of the ketone which was obtained from the *ap*. The other one was assigned to the 1,3-dimethyl compound (**4-133**) from spectroscopic evidence. Here, the ipso attack to the 1-position took place and the 1-methyl group migrated to the 2-position from the benzenonium ion which was produced by the attack.

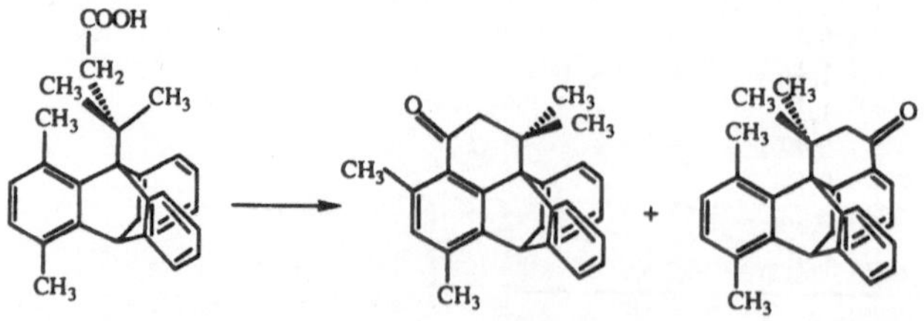

sc-**4-131**   **4-133**   sc-**4-132**

Table 4.16. Effects of solvent and catalyst on the formation
ratios of two ketones (**4-133** and sc-**4-132**)

Solvent	Lewis Acid	sc-**4-132**/**4-133**
CH_2Cl_2	$TiCl_4$	0.96
C_6H_6	$TiCl_4$	0.89
C_6H_{12}[a)]	$TiCl_4$	1.2
C_6H_6	$SnCl_4$	1.4

[a)] cyclohexane.

The formation ratios of the two ketones are again dependent on the solvent
and the Lewis acid used. The results are summarized in Table 4.16. In nonpolar
solvents, the attack on the benzeno bridge that bears no substituent is favored,
whereas the ipso attack on the 1-position is favored in polar solvents in a
relative sense. The stronger Lewis acid, $SnCl_4$, favored the ipso attack. The
results are again in conformity with the tight and loose ion pair model. The
tighter ion pair attacks the benzeno bridge without a substituent due to the
steric effect and the ipso attack becomes easier when the ion pair is loose.

4.2.2.3 Diazotization of Amines

Diazonium salts are known to be sources of carbocations. Therefore, the cation-
forming reaction of rotamers can be examined with this route as well. The
required amine could be obtained via the Curtius reaction of the acid azide
obtained from the acid chloride. Diazotization of the amine was carried out with
the use of isopentyl nitrite and acetic acid in most cases.

The simplest example of the difference in the reactivity of rotamers in this
reaction was provided by a 1,4-dimethoxy compound (**4-134**) [54]. The *ap*
isomer afforded a mixture of olefins (**4-135** ~ **4-137**), a cyclic compound (**4-138**)
which is expected from the attack of the intervening cation on one of the

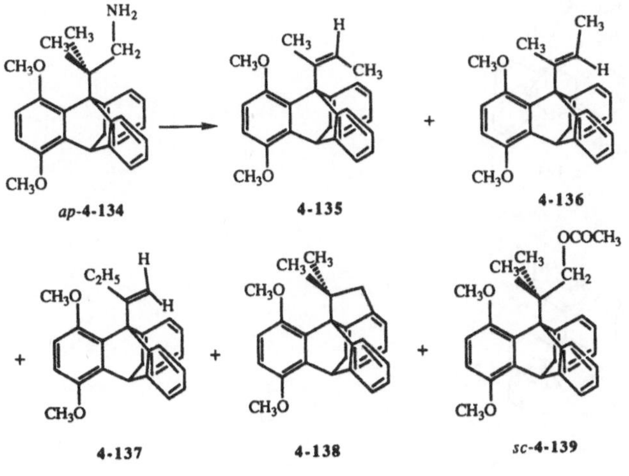

ap-**4-134**   **4-135**   **4-136**

4-137   **4-138**   sc-**4-139**

benzeno bridges in proximity, and an acetate (**4-139**). The reaction mechanisms must involve the Wagner-Meerwein rearrangement as already discussed for the tetrachloro compound. Different from the previously discussed case is the formation of 9-(1-ethylethenyl) compound (**4-137**). The difference is not well explained but the difference in the base present in the system should be responsible. At any rate, the main products are olefins and the cyclic compound and the acetate are formed in only a few percent yields.

The interesting point here is that, though products derived by rearrangement were obtained in ample amounts, there was no product which bears the triptycyl group at the termini of the C_4 unit which is derived from the 9-substituent of the original compound (**4-134**). Diazotization of aliphatic compounds is known to undergo extensive Wagner-Meerwein rearrangement and detailed study on the reaction mechanisms has been carried out. The general conclusion is that alkyl migration takes place simultaneously when the nitrogen molecule departs from the diazonium salt, the migrating group being at the *trans* position of the departing nitrogen [55]. If this mechanism were applied to the present case, compounds which are derived by the triptycyl migration should be obtained, because the diazonium moiety should be anti-parallel to the triptycyl group in this case due to the steric requirement. Namely, in the normal case, a cation **4-141** should be formed from the diazonium salt (**4-140**). Absence of such a product is indicative of that the nitrogen leaving and the alkyl migration in the diazonium decomposition are independent steps, differing from the general concept of this reaction. This is probably caused by the fact that 9-triptycyl cation is not stabilized, as evidenced by a fact that a diazonium salt derived from an 9-aminotriptycene derivative is stable at $-78\,^\circ\text{C}$ [56], while such is not the case for other aliphatic diazonium salts. Thus the cation (**4-142**) formed by denitrogenation from **4-140** rearranges to a more stable cation (**4-143**), where deprotonation takes place to produce the observed olefins.

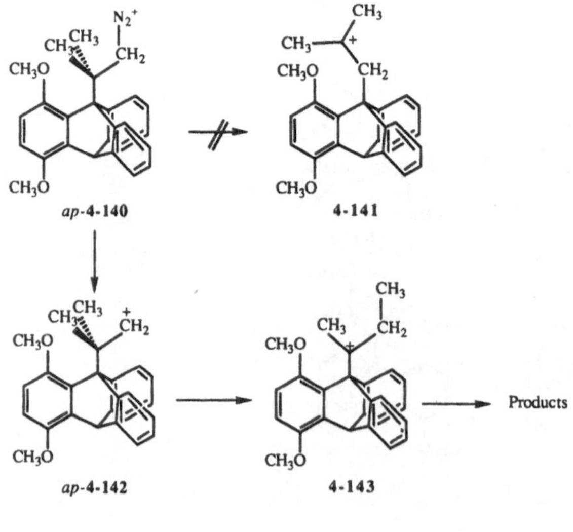

By contrast, the *sc* isomer (**4-134**) gave a product whose spectroscopic data were consistent with the cyclic ether structure (**4-144**), with 9-(2-acetoxy-1,1-dimethylethyl)-1,4-dimethoxytriptycene (*sc*-**4-139**) in 17% yield. No olefinic products were detected.

Formation of this ether was expected, because if a cationic center is formed in proximity of the 1-methoxy group (*sc*-**4-145**), they will react with each other to form an oxonium ion that is then demethylated by the S_N2 type reaction, as found in other instances. The increase in the yield of the acetate with respect to the *ap*-form is of interest. This will mean that the life-time of the intermediate cation is long relative to the case of the *ap*-isomer.

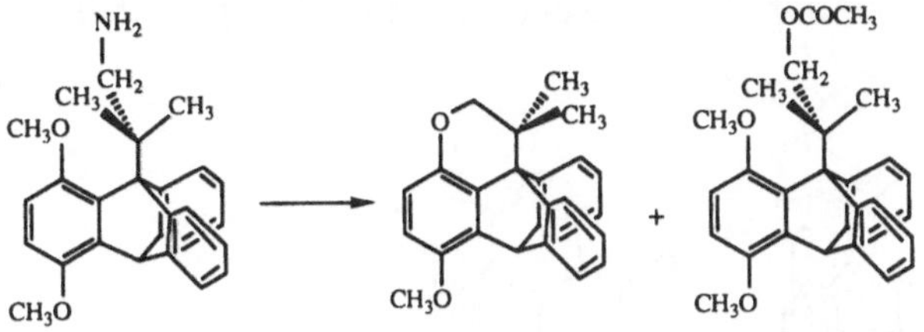

sc-**4-134**   **4-144**   *sc*-**4-139**

The results are consistent with the carbocation formation rather than the simultaneous leaving of nitrogen with the migration of an alkyl group from the diazonium ion (*sc*-**4-140**). The increase in the yield of the *sc*-acetate (*sc*-**4-139**) with respect to the case of the *ap*-acetate (*ap*-**4-139**) will mean that either the cation is stabilized by participation of oxygen (**4-145**) or, less likely, the formation of the oxonium ion (**4-146**) from the oxygen-participated cation (**4-145**) is reversible. At any rate, the oxygen-participation is so strong that it prevents isomerization of the intervening cation. The cause of this facile reaction to form the cyclic ether must be the small distance between the two sites concerned: the distance between the carbocation and the 1-methoxy oxygen is ca. 3.0 Å and that between the carbocation and the benzeno bridge is ca. 3.1 Å.

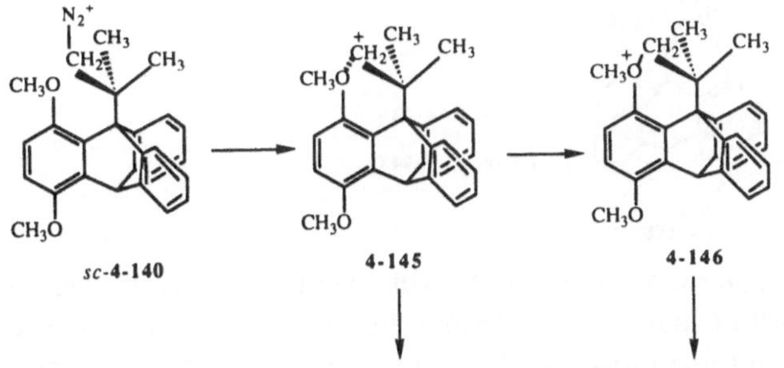

sc-**4-140**   **4-145**   **4-146**

4-144   *sc*-**4-139**

Finding a very strong participation of the 1-methoxy group, one would be interested to see participation of other heteroatoms. To date, participation of 1-chloro substituent is observed. Diazotization of 2-methyl-2-(1,2,3,4-tetrachloro-9-triptycyl)propylamine rotamers (**4-147**) was carried out in essentially the same way as for other amines [57]. The products from the *ap*-amine are olefins, a cyclized compound, and an acetate as were observed in the *ap*-dimethoxy compound. However, when the *sc*-amine was diazotized, there was a main product, which was identified as 2-methyl-2-(1,2,3,4-tetrachloro-9-triptycyl)propyl acetate (**4-150**), together with minor amounts of hydrocarbons (**4-124**, **4-125**, **4-148**, and **4-149**).

The formation of the acetate in an ample amount must be attributed to the longer life of the carbocation (*sc*-**4-120**), produced from the *sc*-amine, because of the participation of the 1-substituent, with respect to that (*ap*-**4-120**) formed the *ap*-amine. Thus there is a competition between the isomerization to produce another cation (**4-121**) and the attack of the acetate ion on the cation. The results indicate that the formation of the acetate (*sc*-**4-150**) is the preferred process for *sc*-**4-120**.

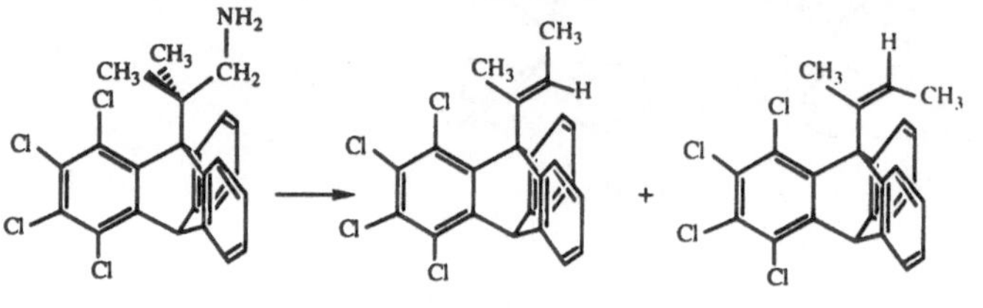

ap-4-147   4-124   4-125

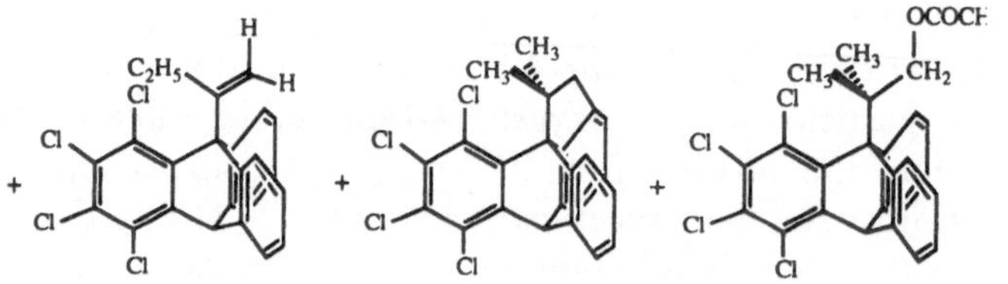

4-148   4-149   ap-4-150

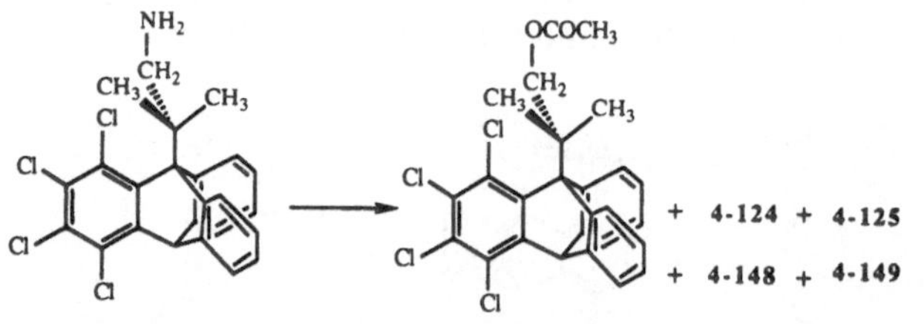

sc-4-147   sc-4-150   + 4-124 + 4-125 + 4-148 + 4-149

In accord with this postulate, when trifluoroacetic acid was used in place of acetic acid, the yield of trifluoroacetate, corresponding to compound *sc*-**4-150**, was diminished to 20% and formation of the olefins and the cyclized product increased as judged from the spectra. This change in the product formation is derived by the fact that the trifluoroacetate anion is inferior in nucleophilicity. Namely, although the intervening carbocation is stabilized by chlorine-participation, the nucleophilicity of trifluoroacetate is so low that the reaction to form the trifluoroacetate is slow to allow the rearrangement of the carbocation to occur.

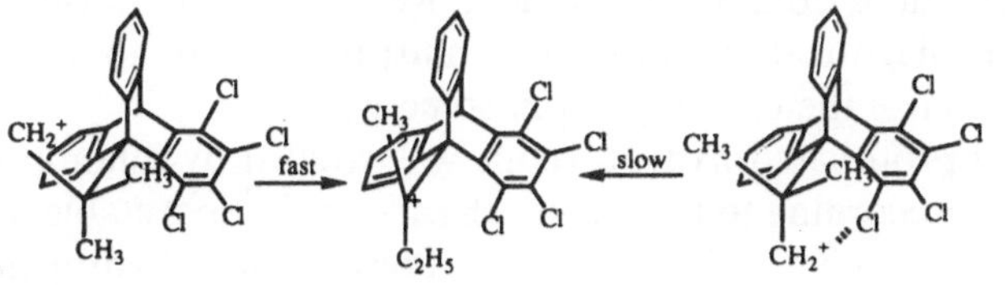

ap-4-120   4-121   sc-4-120

Diazotization of 2-methyl-2-(1,4-dimethyl-9-triptycyl)propylamine rotamers
(4-151) revealed another interesting aspect. As already experienced in other
compounds, the *ap* form afforded olefins (4-152 to 4-154) and the cyclized
compound (4-155) in much the same way, in addition to ca. 3% acetate (*ap*-4-
156). By contrast, diazotization of the *sc*-amine in benzene in the presence of
acetic acid afforded a compound which was identified by independent synthesis
as the carbocation-insertion product to the 1-methyl group (4-157). The corresponding acetate (*sc*-4-156) was also found more abundantly (ca. 16%) than the
ap case among the products which contained olefins (4-152 ~ 4-154) and the
cyclized compound (4-155), as were obtained from the *ap* [58].

The insertion of the carbocation to an alkyl group is rather a rare case,
though it is commonly observed in super acid media [59], and cyclopropane
ring formation both in nature [60] and in laboratories [61] is well known. It is

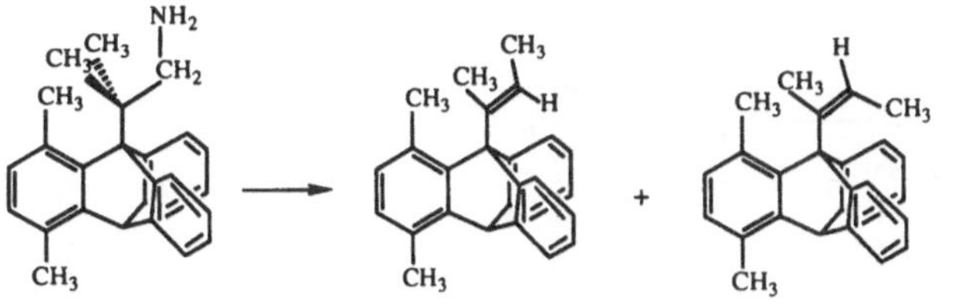

ap-4-151   4-152   4-153

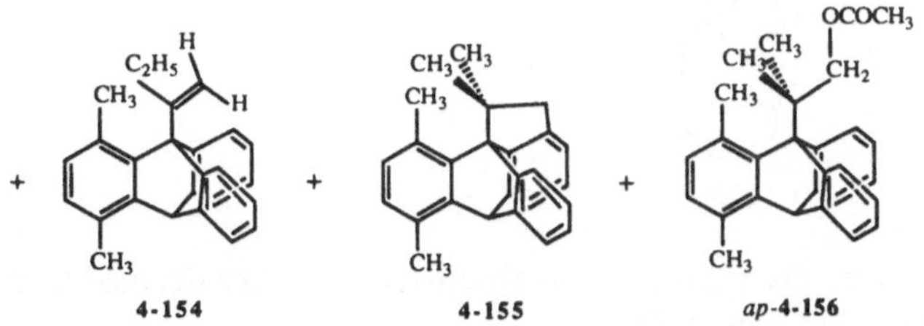

4-154   4-155   ap-4-156

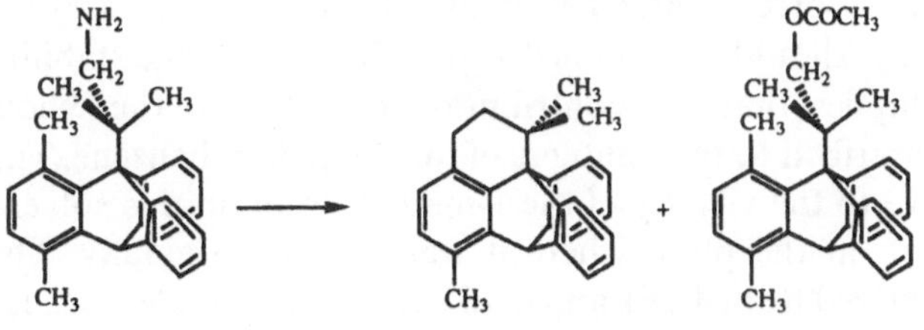

sc-4-151   4-157   sc-4-156

+   Others

also interesting that the acetate is obtained from the *sc*-isomer up to 16% yield, though it is only 3% in the diazotization of the *ap*. It must be assumed that the 1-methyl group acts to stabilize the carbocation in some ways.

The stable structure of protonated hydrocarbons was studied by quantum chemical calculations [62]. According to the results, the species is most stable in a form which is close to a coordinated form of the carbocation to a hydrogen molecule. If this structure is applied here, they may be written as follows. Although it is stated that this form can give carbocation-insertion products easily, it is not easy to understand why this structure gives the acetate as well. Accordingly, it is postulated that the most stable forms are in equilibrium to which all of the possible coordinated forms contribute: a proton coordinated to a C–C bond (**4-158**), a proton coordinated to a C–H sigma bond (**4-159** and **4-160**), a carbocation coordinated to a hydrogen molecule (**4-161** and **4-162**), and a carbocation coordinated to a C–H sigma bond (**4-163** and **4-164**). It will be natural to consider that these species interchange without a high barrier.

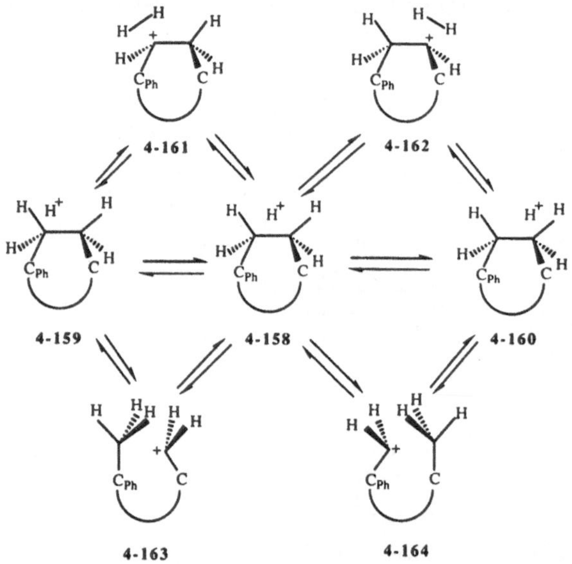

One point which deserves discussion is the absence of acetates derived from cation **4-164** and cations **4-161** and **4-162**.

If the above postulate is correct, the compound, which carries an acetoxy group at the 1-methyl group, should be obtained, especially because the stability of the benzylic cation implies that the contribution of **4-164** to the whole equilibrium is great. It is ascribed to the tightness of the ion pair in benzene. The acetate anion cannot move to the vicinity of the 1-methyl cation in this solvent and reacts with the cation in the place where it was formed originally. The absence of the products derived from **4-161** and **4-162** seems reasonable, because if those were to be formed, the acetate anion would have attacked from the rear side of the cations and that is blocked by the ring.

In the hope of finding 1-substituted methyl compounds, various solvents were used for diazotization of the sc-amine (sc-**4-151**). Unfortunately, however, no acetate was found under these conditions, even the compound (sc-**4-156**) which carries the acetoxy group in the 9-substituent was not found. The major product was the cyclized compound, the yield being ca. 55%.

The cause of this solvent effect is not well understood at present, but the following may be possibilities. In a polar solvent, ion pairs are looser and are capable of insertion to the C–C bond more easily in a polar solvent than in a nonpolar. It is also possible that the protonated hydrocarbon is more stable than the hydrogen-carbocation-coordinated species due to the better solvation in polar solvents.

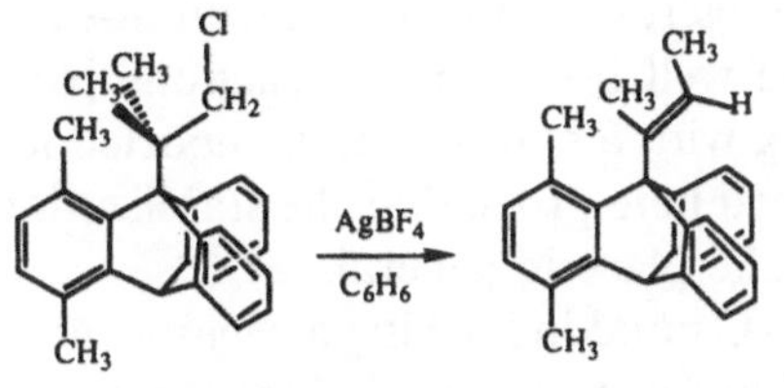

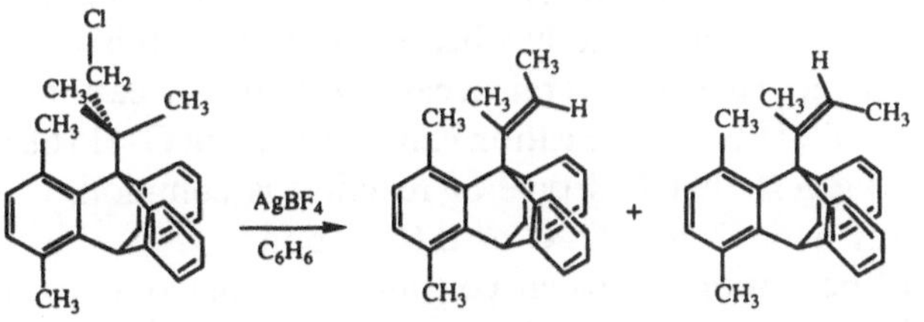

Treatment of the rotameric 1,4-dimethyl-9-(2-chloro-1,1-dimethylethyl)-triptycenes (**4-165**) with silver tetrafluoroborate proceeded in much the same way except two points. Namely, the ap-isomer afforded only one olefin (**4-152**) and the sc-isomer failed to give the five-membered ring compound (**4-155**) [63]. The Z-olefin (**4-152**) and the six-membered ring compound (**4-157**) were the major products.

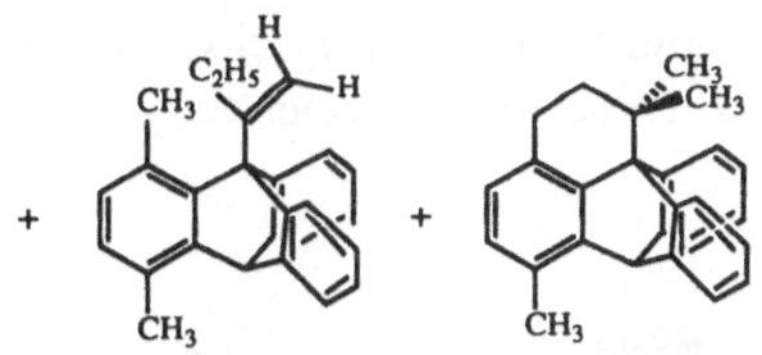

Silver salt-assisted ionization of 9-tert-butyl-1-chloromethyl-4-methyltripty-cene (**4-166**) was tried in benzene. A preliminary result indicated that indeed the cyclization reaction occurred to some extent to give compound **4-157** to prove that the cation produced at the 1-methyl group can insert to the C–H bond of the tert-butyl group [63]. This result supports the above postulate.

4-166   4-157

The results clearly show that a methyl group should be considered to participate in stabilizing a carbocation. The degree of stabilization may be comparable with that of a chloro substituent, because, under the same conditions, the 1-chloro compound (sc-4-147) gave olefins and the acetate (sc-4-150) in similar yields. Hydride transfer reactions across the ring in medium-sized rings [64] is known and hydride shifts are well documented reactions [65]. Stabilization of two carbocations by binding with a hydride ion is reported for medium sized rings [66]. The results described here prove that the stabilization of a carbocation with an alkyl group can take place in general.

Insertion of carbocations to a C–H or C–C bond in forming a cyclopropane ring or in super-acid solution may be considered a special case. However, a case of carbocation insertion to a C–H bond to form a four-membered ring was reported as early as 1960 [67]. Another case has been described in this book (page 131). It has also been reported in several occasions that if a cation is formed in close proximity of a C–H bond, the cation can insert to the C–H [68]. The results described here suggest that this type of reaction is common, if a carbocation is located closely enough to a C–C or a C–H bond.

The generalization may be important when considering a reaction path. Traditionally a hydrocarbon molecule or a hydrocarbon group has been considered to exhibit only repulsive forces when an cationic species approaches it. It is now necessary to consider that, at a very close distance such as that in a transition state of a reaction, these ionic species may exhibit attractive forces with hydrocarbons.

4.2.3 Other Types of Reactions

4.2.3.1 Reactions of Chlorides with *tert*-Butyllithium

Treatment of alkyl halides with butyllithium is known to proceed with single electron transfer, if the halogen is bromine, whereas no radical intermediate is detected if the halogen is iodine in the same reaction [69]. Chlorides did not react in the same treatment.

Only one example of such a treatment is known [70], although it is expected that, if a compound carries a ligating atom in the vicinity of the anion forming site, anion formation should be facilitated. The example is 9-(2-chloro-1,1-dimethylethyl)-1,4-dimethyltriptycene rotamers (4-165). Each of the rotamers was treated with *tert*-butyllithium in benzene-hexane and then quenched by

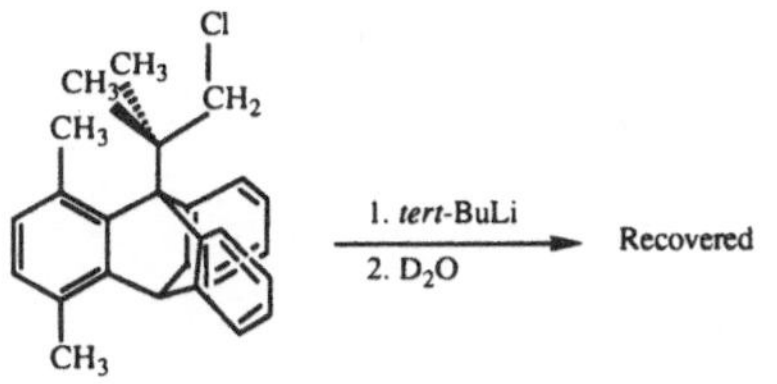

D$_2$O. The result was the recovery of the *ap*-form, whereas the *sc*-form afforded the 1-deuteriomethyl compound (**4-167**) in addition to other compounds [70].

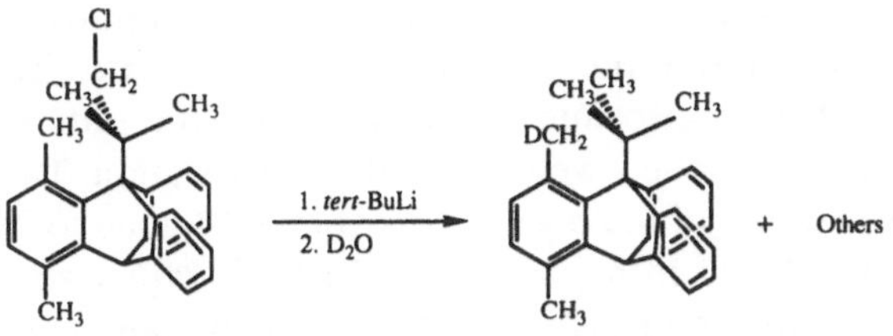

Examination of other products revealed that they were non-deuterated 9-*tert*-butyl-1,4-dimethyltriptycene (**4-168**), 9-*tert*-butyl-1-*tert*-butylmethyl-4-methyltriptycene (**4-169**), and bis(9-*tert*-butyl-4-methyl-1-triptycylmethyl) (**4-117**). These products clearly indicate that the reaction proceeded with a radical path way.

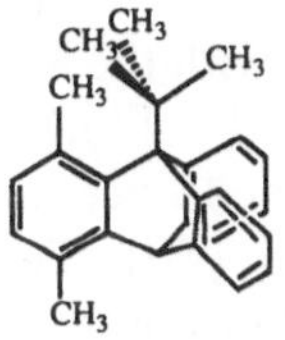

The results are interesting in two ways. One is that even chloro compounds reacted with *tert*-butyllithium if it is the *sc*-form. Another point is that, whereas products originated from the 1-radicals (**4-119**) were obtained, no products derived from the radical center in the 9-substituent (**4-118**) was found, although compound **4-168** could be such a product.

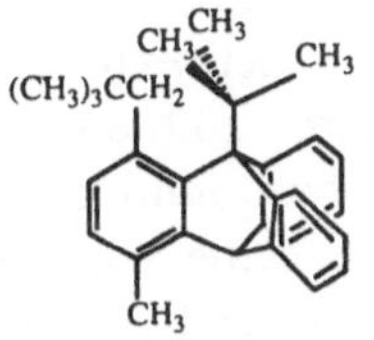

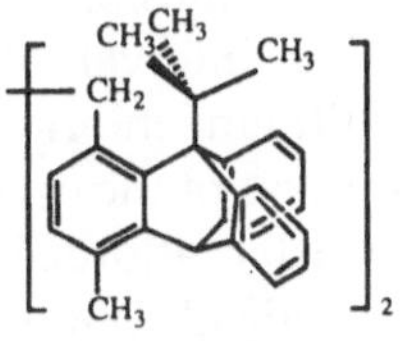

The first point implies that alkyl chlorides do react with *tert*-butyllithium but the rates are rather low. In the case of the *sc*-isomer, due to the presence of a benzylic methyl group, hydrogen transfer is facile. Thus as soon as the single electron transfer occurs at the CH$_2$–Cl group in the *sc*-isomer, the hydrogen in the 1-methyl group seems to migrate to form a benzylic radical (**4-119**). In the case of *ap*-isomer, it seems that the single electron transfer does not take place to an observable extent, because there is no factor to enhance the reactivity of the CH$_2$–Cl group in *ap*-**4-165**.

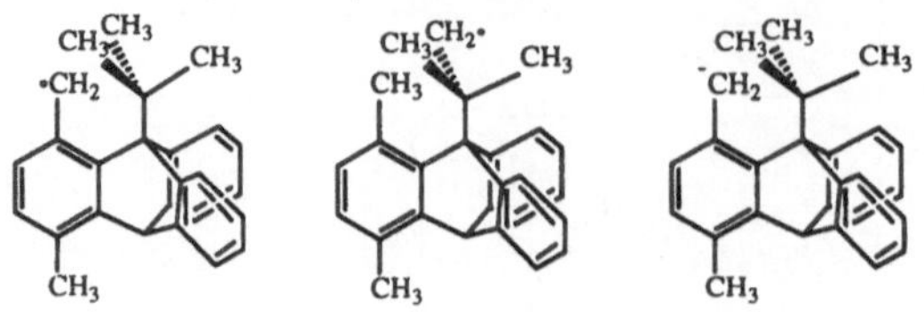

4-119   4-118   4-170

The second point makes a contrast to the case of peroxy ester decomposition (see page 139). The difference may imply the difference in intermediates but details are not known at present. It is also interesting that both the radical-originated and the anion-originated products were isolated in the reaction. This will be the result reflecting the stability of the benzyl radical (4-119). Due to its stability, the second electron transfer to form the anion (4-170) is rather slow and 4-119 survives to react with other radicals or to abstract hydrogen.

4.2.3.2 Chlorodecarboxylation

Hunsdiecker reactions of rotameric 3-methyl-3-(1,4,-dimethyl-9-triptycyl)-butanoic acid (4-131) in the presence of lithium chloride has been reported [71]. The ap-isomer afforded the corresponding chloride (4-165) and a cyclized compound (4-155) in 15 and 70% yields, respectively. The formation of the cyclic compound is not an artifact, i.e. it is not a secondary product which could be formed from the chloride (4-165) if the lead salt present in the system acts as an Lewis acid. This was proved by the constant formation ratios of the chloride to the cyclic compound at various intervals of the reaction together with the fact that increasing the amount of lithium chloride increased the yield of the chloride relative to the cycle. The formation of the cyclic compound is expected since the decomposition of the peroxyester afforded the same compound [50]. The Hunsdiecker reaction, being a radical type reaction [35], is expected to proceed in a similar fashion.

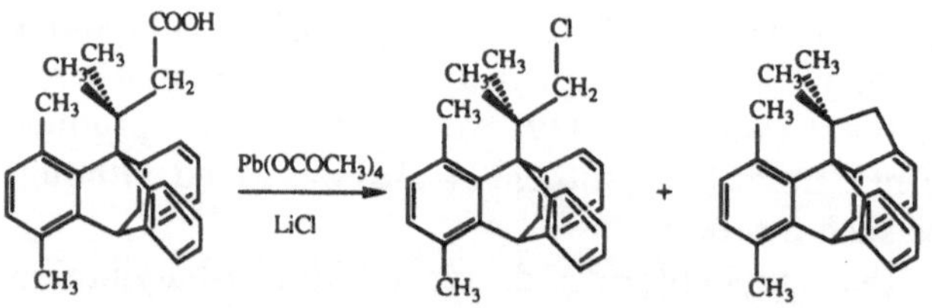

ap-4-131   ap-4-165   4-155

Also interesting is the facile formation of the cyclic compound (4-155) in this reaction. Under normal conditions, the Hunsdiecker reaction in the presence of lithium chloride affords chlorides only: no aromatic substitution has been reported, though it is possible to find products due to this reaction in the absence of lithium chloride. This is attributed to the very short distance between the radical center and the aromatic ring in this system. It also means that, although the formation of chlorides under the Hunsdiecker reactions is attributed to ligand coupling, the alkyl radical has ample chance to dissociate from the lead center to react with the aromatic ring.

The involvement of the free radical species in the Hunsdiecker reaction in the presence of lithium chloride becomes more convincing when one sees the results of the reaction of the *sc*-isomer (**4-131**). It gives 9-*tert*-butyl-1-chloromethyl-4-methyltriptycene (**4-166**) and 1-acetoxymethyl-9-*tert*-butyl-4-methyltriptycene (**4-171**) in 35 and 50% yields, respectively. No cyclic compound was detected among the products.

the radical transfer from the substituent (**4-118**) where the radical is originally formed to the 1-position (**4-119**) is very fast. It is even faster than the cyclization to the benzeno bridges, the expected product (**4-113**: X=CH$_3$, Y=H) being not found, as was the case for the decomposition of the corresponding perester (*sc*-**4-114**).

Secondly, the acetoxy derivative, that was not detected in the case of *ap*, was obtained in the larger amount in the case of the *sc*. This might be partly due to the stability of the benzylic radical. The stable radical can survive without reacting with the chlorine species, until it is oxidized by the lead(IV) salt, the reaction of which is known to be slow. It is also possible that the steric protection of the 9-substituent renders the life time of the 1-methyl radical (**4-119**) long. Although there is still another possibility that the benzyl chloride type compound reacts with the lead(IV) or lead(II) acetate in the system, the long life of the benzylic radical (**4-119**) seems to be the main factor for the formation of the acetate, because radical coupling products (**4-117** and **4-169**) are obtained in decomposition of the perester (*sc*-**4-109**) and in lithiation of the halide (*sc*-**4-165**).

4.2.3.3 Bromine Addition Reactions

Another case of investigation on the reactions of rotamers is bromination of 9-allyl-1,2,3,4-tetrachlorotriptycene (**4-172**) [8]. This compound, differing from the cases discussed earlier in this chapter, could exist as a mixture of *ap* and ± *sc* rotamers that were not separable at room temperature due to low barriers to rotation. Although examination of its ^{1}H NMR spectra at room and lower temperature indicates that it exists as the *ap* isomer exclusively, the barrier to rotation, ca. 12 kcal/mol, is low enough to enable the substrate or an intermediate of the reaction to rotate during the reaction at room temperature.

Bromination of this compound in chloroform afforded a mixture of the corresponding dibromide (**4-173**) and *cis* and *trans*-bromo-olefins (**4-174**), the latters being the major products. The formation of the bromo-olefins or rearranged compounds in bromination of congested olefins is known [7]. Therefore, the results are consistent with the congested steric environment of the reaction center in this molecule.

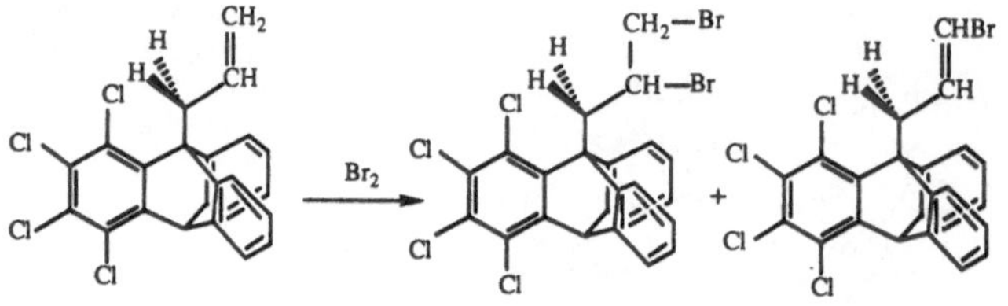

The problem is "in which conformation does the compound react?". The reaction can occur exclusively in the *sc* conformation, if the reactivity of the *sc*-isomer exceeds that of the *ap* to an extent of several kcal/mol in energy barrier for the reaction, though its population is scarce. Therefore, the consideration of the reaction path should include these factors. This kind of discussion is now possible since we know the reactivity of rotational isomers to some extent.

The conformation about the bond connecting the vinyl and the methylene groups in the 9-substituent is considered to be most stable when the vinyl group takes the position upright to the triptycene skeleton for the following reasons, though propene is known to be stable in the methyl–CH eclipsing conformation [72]. Firstly, if the vinyl group in the compound takes the eclipsing conformation with one of the CH bonds, the =CH group in the vinyl is very close to one of the benzeno bridges. This must be unstable because of large steric repulsion. For the same reason, the vinyl group cannot take eclipsing conformation with the CH_2–C_9 bond. Secondly, of the two possible conformations, vinyl group-inside (*ap*-**4-172b**) and outside (*ap*-**4-172a**), after establishing the stability of the non-eclipsing conformation about the CH_2–vinyl bond, the inside conformation must be prohibitively unstable again due to the steric effect. Therefore, the *ap*

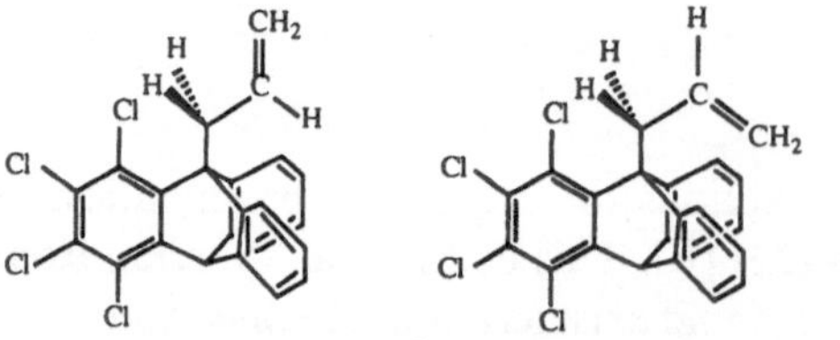

conformation with respect to the vinyl group and the tetrachlorobenzeno moiety about the CH_2–C_9 bond is considered to be the most stable. This type of conformation is known to be stable also in an aldehyde which possesses an oxygen instead of the vinylic CH_2 group in *ap*-**4-172** [73].

In order to diagnose the cause of the bromo-olefin formation, bromination of various related triptycenes was examined.

Interestingly, 9-allyltriptycene (**4-175**: X=Y=H), 9-allyl-2,3-dichloro-triptycene (**4-175**: X=H, Y=Cl), and 9-allyl-1,4-dimethyltriptycene (**4-175**: X=CH$_3$, Y=H) all afforded the addition compound (**4-176**) exclusively, when treated with bromine under the same conditions as for the tetrachloro compound. These results suggest that the bromo-olefin formation is not caused by the steric effect or the inductive effect of the substituent.

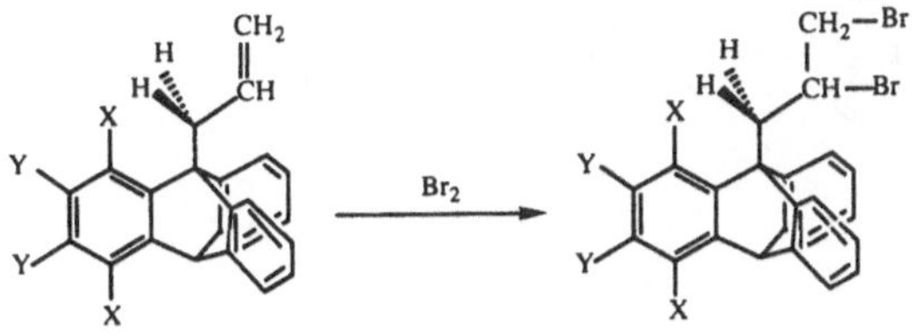

4-175   4-176

If the steric effects were the main cause for the bromo-olefin formation, then the 1,4-dimethyl compound in the *sc* form should have given a corresponding bromo-olefin. If the inductive effects were important factors for giving the bromo-olefin, the 2,3-dichloro compound should have given a detectable amount of the bromo-olefin. These considerations lead to the conclusion that the bromo-olefin is not formed in the *ap* conformation but in the *sc*. Then the participation of the 1-chloro substituent can be the reason for this bromo-olefin formation.

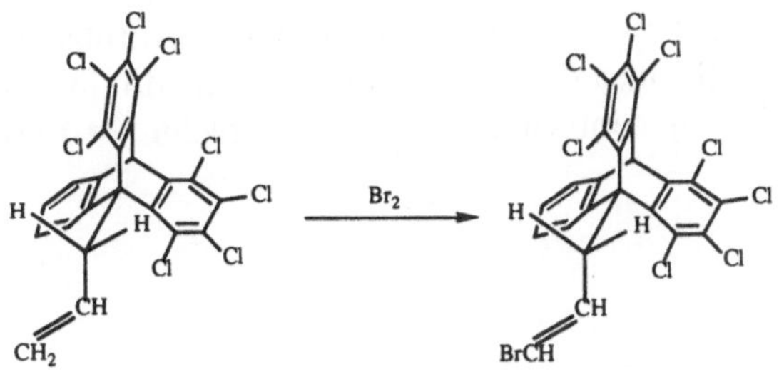

sc-**4-177**   sc-**4-178**

Bromination of 9-allyl-1,2,3,4,5,6,7,8-octachlorotriptycene (**4-177**) under the same conditions as the others afforded bromo-olefin (**4-178**) only: no dibromo compound was detected. Since this compound gives an AB signal only for the methylene protons in its ^{1}H NMR spectra at room temperature, it should exist

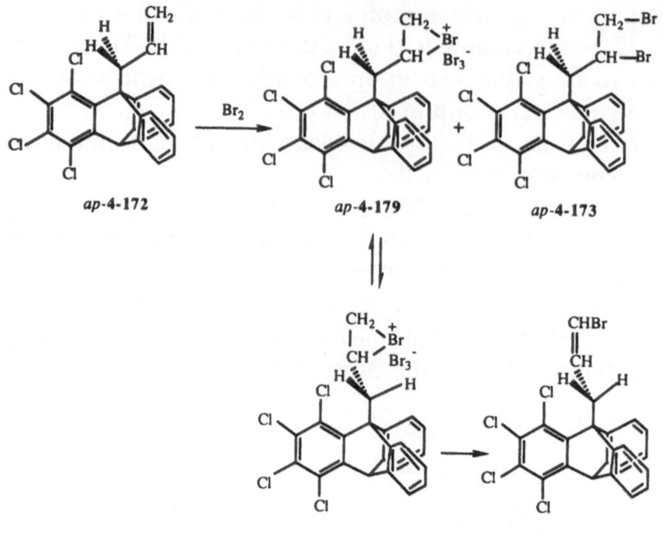

ap-4-172   ap-4-179   ap-4-173

sc-4-179   sc-4-174

as *sc* conformers exclusively. This is reasonable because the *ap* conformation is unstable due to the steric effects that are given by the two flanking 1- and 8-chloro substituents. These results and considerations of the structures of the species involved lead to the conclusion that the bromo-olefin (**4-178**) is formed from bromine-bridged intermediate cations (*sc*-**4-179**) that are stabilized by the chlorine participation.

These brominations are rather slow reactions with respect to the ordinary bromination of olefins. The results are attributed to the steric effect of the triptycene systems that carry substituent(s) in the vicinity of the reaction center because a sterically protected, bridged bromonium ion derived from bisadamantylidene is known to be stable [74]. Competitive reactions indicate, however, the octachloro compound (**4-177**) and 9-allyl-1,4,5,8-tetramethyltriptycene (**4-180**) show only negligible reactivity toward bromine with respect to unsubstituted 9-allyltriptycene. The results support the idea that the bromine-cation addition to the olefin takes place in the *ap*-conformation of the olefin (**4-175**) but not in the *sc*.

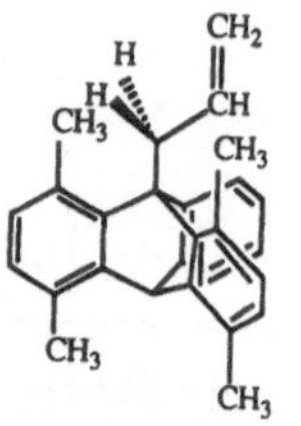

4-180

Then the formation of the bromo-olefin in the bromination of the tetra-chloro compound must take place via the following steps: bromine-cation addition to the olefin in the *ap*-conformation to form a bridged bromonium ion (*ap*-**4-179**), internal rotation to the *sc*-position (*sc*-**4-179**) in the bridged bromon-ium ion, and then deprotonation. The *sc*-olefin (*sc*-**4-174**) thus formed isomerize to the more stable *ap*-**4-174** quickly even at a low temperature. The next problem is to decide when the dibromide is formed from the bridged bromon-ium ion, in the *ap* conformation or in the *sc*.

The keys to answer this question have already been mentioned. The octachloro compound failed to give any of the dibromide. This is reasonable on steric grounds. The proton-abstracting species must penetrate deeply into the triptycene skeleton in order to form the olefin and this trespassing is unfavorable especially when a substituent is present at the *peri*-position. Thus the dibromide

Table 4.17. Formation ratios of the bromo-olefins vs the adduct in the bromination of **4-172** at various temperatures

Temperature/°C	Ratio(substitution/addition)
40.7	1.54
20.0	2.40
0.0	3.09
− 17.0	7.36

is formed in the *ap* conformation. This implies that the life-time of the bridged bromonium ion is longer than the time needed for internal rotation.

Then it becomes interesting to see how the formation ratios of the bromo-olefin and the dibromide change at various temperatures. This can be studied because the dibromide formation and the internal rotation can both be con-sidered unimolecular: the former is considered unimolecular because the inter-mediate exist as ion pairs and the rate-limiting step for the formation of the dibromide is not the formation of the bromonium ion but the collapse of the ion pair.

Examination of the formation ratio, bromo-olefin/dibromide, at various temperatures produced the results shown in Table 4.17. The ratio increases as temperature is decreased. The Arrhenius plot yields the difference in activation energy as 3.0 kcal/mol and that in entropy of activation as 8.0 e.u. If we assume the barrier to internal rotation in the bridged bromonium ion to be the same with a model compound, 9-benzyl-1,2,3,4-tetrachlorotriptycene, then the activa-tion energy for bromide formation is ca. 16 kcal/mol. Using these values, one estimates the half-life of the bridged bromonium ion to be ca 0.1 s at 0 °C.

4.3 References

1. Ōki M, Tsukahara J, Moriyama K, Nakamura N (1987) Bull Chem Soc Jpn 60: 223
2. Moriyama K, Ōki M (unpublished work)

3. Ōki M, Tsukahara J, Sonoda Y, Moriyama K, Nakamura N (1989) Bull Chem Soc Jpn 62: 621
4. Fahey RC, Schneider HJ (1968) J Am Chem Soc 90: 4429. Rolston JH, Yates K (1969) J Am Chem Soc 91: 1467
5. Fahey RC, Schubert C (1965) J Am Chem Soc 87: 5172. Yates K, Leung HW (1980) J Org Chem 45: 1401
6. Ōki M, Otake K, Shionoiri K, Ono M, Toyota S (1991) Chem Lett 597
7. Allinger NL, Tushaus LA (1967) Tetrahedron 23: 2051. Burgoine KT, Davies SG, Peagram MJ, Whitham GH (1974) J Chem Soc, Perkin I 2629
8. Hatakeyama S, Mitsuhashi T, Ōki M (1980) Bull Chem Soc Jpn 53: 731
-9. Ōki M, Ohira M (1984) Bull Chem Soc Jpn 57: 3025
10. Saito R, Ōki M (1982) Bull Chem Soc Jpn 55: 3267
11. Saito R, Ōki M (1982) Bull Chem Soc Jpn 55: 3273
12. Sonoda Y, Tsukahara J, Nakamura N, Ōki M (1989) Bull Chem Soc Jpn 62: 621
13. Peterson DJ (1968) J Org Chem 33: 780. Hudrlik PF, Peterson D (1975) J Am Chem Soc 97: 1464
14. Rocek J (1966) Oxidation of aldehydes by transition metals. In: Patai S (ed) The chemistry of the carbonyl group, Wiley, New York, p 462
15. Mowry DT (1948) Chem Rev 42: 189. Mukaiyama T, Tonooka K, Inoue K (1961) J Org Chem 26: 2202
16. Nakamura M, Ōki M (1974) Tetrahedron Lett 505
17. Murata S, Kanno S, Tanabe Y, Nakamura M, Ōki M (1984) Bull Chem Soc Jpn 57: 525
18. Murata S, Kanno S, Tanabe Y, Nakamura M, Ōki M (1982) Bull Chem Soc Jpn 55: 1522
19. Bentley TW, Bowen CT, Parker W, Watt CIF (1979) J Am Chem Soc 101: 2486
20. Cram DJ (1953) J Am Chem Soc 75: 332
21. Tichy M, Jonas J, Sicher J (1959) Collect Czechoslov Chem Commun 24: 3434
22. Moriyama K, Nakamura N, Nakamura M, Ōki M (1987) Gazz Chim Ital 117: 655
23. Nakamura M, Ōki M (1987) Bull Chem Soc Jpn 60: 673
24. Ōki M, Iwamura H (1961) Bull Chem Soc Jpn 34: 1395
25. Nakamura M, Ōki M (1986) Chem Lett 1363
26. Nakamura M, Nakamura N, Ōki M (1977) Bull Chem Soc Jpn 50: 1097
27. Margerison D, Newport JP (1963) Trans Farady Soc 59: 2058
28. Nakamura M, Ōki M (1975) Chem Lett 671
29. Brown TL (1966) Adv Organomet Chem 3: 365
30. Nakamura M, Suzuki Y, Ōki M (1985) Bull Chem Soc Jpn 58: 2370
31. Ho TL (1977) Hard and soft acids and bases principle in organic chemistry, Academic, New York
32. Xu WX, Smid J (1984) J Am Chem Soc 106: 3790
33. Moriyama K, Nakamura M, Nakamura N, Ōki M (1989) Bull Chem Soc Jpn 62: 485
34. Yamada T, Kaneko K, Ōki M (unpublished work)
35. Sheldon RA, Kochi JK (1972) Org React 19: 279
36. Yamada T, Ōki M (unpublished work)
37. Murata S, Sugawara T, Iwamura H (1985) J Am Chem Soc 107: 6317
38. Iddon B, Meth-Cohn O, Scriven EFV, Suschitzky H, Gallagher PT (1979) Angew Chem Int Ed Engl 18: 900
39. Sonoda Y, Iwamura H (private communication)
40. Cristol SJ, Pennelle DK (1970) J Org Chem 35: 2357
41. Yamamoto G, Ōki M (1987) Chem Lett 2181
42. Yamamoto G, Ōki M (1987) Chem Lett 1163
43. Seki S, Morinaga T, Kikuchi H, Mitsuhashi T, Yamamoto G, Ōki M (1981) Bull Chem Soc Jpn 54: 1465
44. Bentrude WG, Martin JC (1964) J Am Chem Soc 84: 1561
45. Mikami M, Toriumi K, Konno M, Saito Y (1975) Acta Crystallogr, Sect B, 31: 2474
46. Yonemoto K, Kakizaki F, Yamamoto G, Nakamura N, Ōki M (1985) Bull Chem Soc Jpn 58: 3346
47. Kuivila HG (1968) Acc Chem Res 1: 299
48. Kurhanewicz J, Jurch GR Jr. (1987) J Am Chem Soc 109: 5038 and papers cited therein.
49. Ōki M, Endo M, Noda Y, Toyota S, Yamasaki M, Shibahara T (1992) Chem Lett 273
50. Tanaka Y, Tanuma T, Ōki M (unpublished work)
51. Kikuchi H, Seki S, Yamamoto G, Mitsuhashi T, Ōki M (1982) Bull Chem Soc Jpn 55: 1514

52. Ōki M, Tanuma T, Tanaka Y, Yamamoto G (1988) Bull Chem Soc Jpn 61: 4309
53. Tanaka T, Yonemoto K, Nakai Y, Yamamoto G, Ōki M (1988) Bull Chem Soc Jpn 61: 3239
54. Ōki M, Taguchi Y, Toyota S, Tanaka T, Yamamoto G (unpublished work)
55. Friedman L (1970) Carbonium ion formation from diazonium ions. In: Olah G, Schleyer P von R (eds), Carbonium ions, vol. II, Wiley, New York, p 655
56. Curtin DY, Klanderman BH, Tavares DF (1962) J Org Chem 27: 2709
57. Tanaka Y, Yamamoto G, Ōki M (1989) Chem Lett 2019 and unpublished works
58. Ōki M, Taguchi Y, Toyota S, Yonemoto K, Yamamoto G (1990) Chem Lett 2209
59. Olah GA, Surya Prakash GK, Sommer J (1985) Super acids, Wiley, New York, chap 3
60. Torssell KBG (1983) Natural product chemistry, Wiley, New York, chap 3
61. Bayless J, Friedman L, Smith JA, Cook FB, Shechter H (1965) J Am Chem Soc 87: 661
62. Lathan WA, Hehre WJ, Pople JA (1971) J Am Chem Soc 93: 808. Bishof PK, Dewar MJS (1975) J Am Chem Soc 97: 2278
63. Ōki M, Taguchi Y, Toyota S (unpublished work)
64. Urech HJ, Prelog V (1957) Helv Chim Acta 40: 477
65. Ahlberg P, Jonsall G, Engdahl C (1983) Adv Phys Org Chem 19: 223. Watt CIF (1988) Adv Phys Org Chem 24: 57
66. Kirchen RP, Okazawa N, Ranganayakulu K, Rauk A, Sorensen TS (1981) J Am Chem Soc 103: 597
67. de Vries L, Winstein S (1960) J Am Chem Soc 82: 5363
68. Biali SE, Rappoport Z (1986) J Org Chem 51: 964 and papers cited therein
69. Ashby EC, Pham TN (1987) J Org Chem 52: 1291
70. Ōki M, Taguchi Y, Toyota S (unpublished work)
71. Ōki M, Okamoto T, Toyota S, Yonemoto K, Yamamoto G (1990) Chem Lett 199
72. Lide DR Jr, Christensen D (1961) J Chem Phys 35: 1374
73. Ōki M, Izumi G, Yamamoto G, Nakamura N (1982) Bull Chem Soc Jpn 55: 159
74. Strating J, Wieringa JH, Wynberg H (1969) J Chem Soc Chem Commun 907

Subject Index